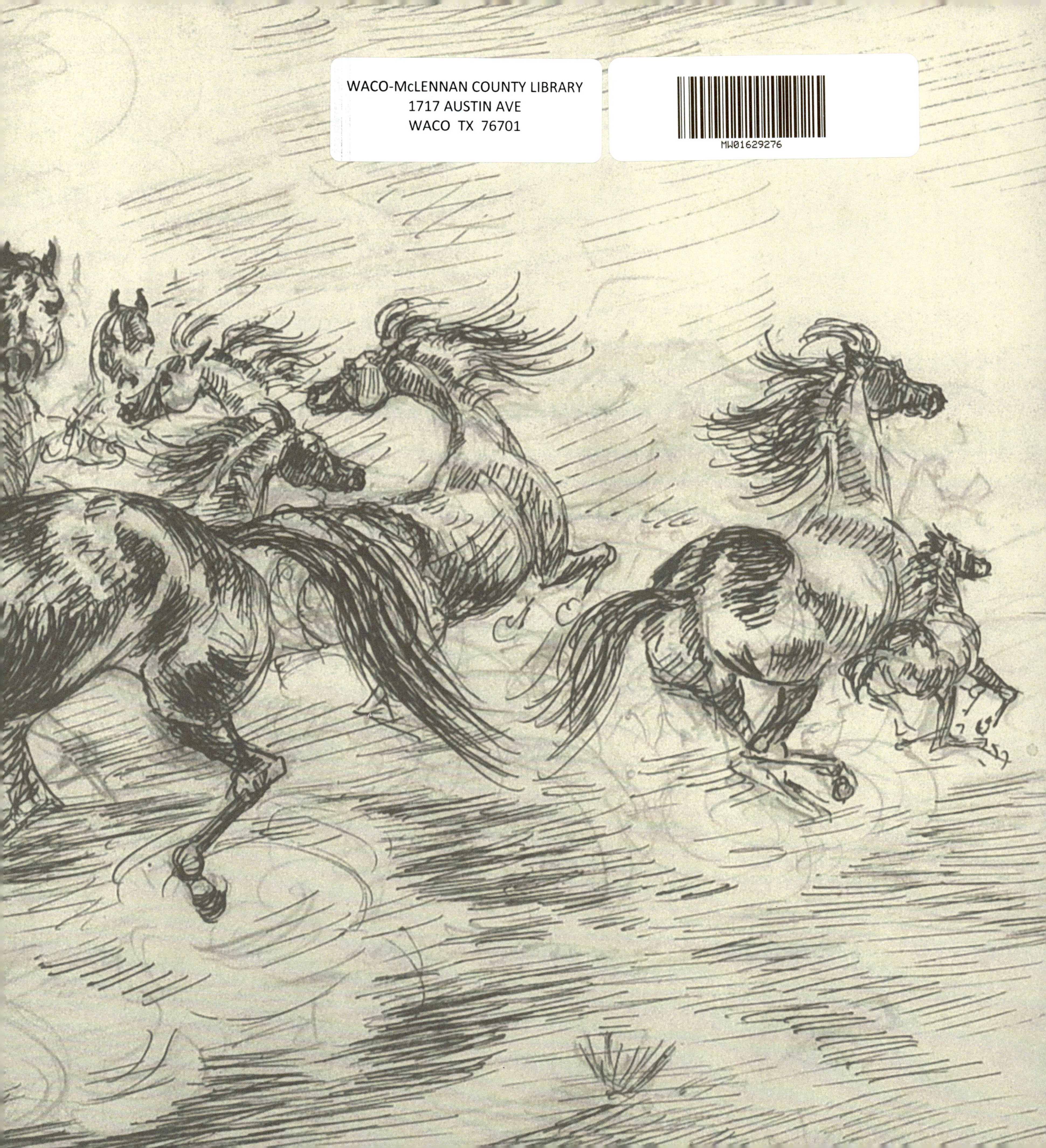

BEASLEY'S VAQUEROS

Beasley's Vaqueros

THE MEMOIRS, ART, AND POEMS OF RICARDO M. BEASLEY

By Andrés Tijerina

FOREWORD BY RON TYLER

Texas State Historical Association

AUSTIN

LIBRARY OF CONGRESS CATALOGING-IN-PUBLICATION DATA

Names: Tijerina, Andrés, compiler. | Beasley, Ricardo M., 1908-1994.
    Works. Selections.
Title: Beasley's vaqueros : The memoirs, art, and poems of Ricardo M.
    Beasley / by Andrés Tijerina.
Description: Austin, Texas : Texas State Historical Association, [2022] |
    Includes bibliographical references and index. | Some Spanish text. |
    Summary: "Beasley's Vaqueros presents the life and work of South
    Texas artist Ricardo M. Beasley, whose vibrant pen-and-ink sketches
    captured the world of vaquero life in South Texas. Despite having lived
    much of his life after World War II, Beasley's art and words capture
    a world in which people and events from decades before his time are
    just as immediate-perhaps even more so-than events of the present
    day. More than just a testament to the talents of a singular, self-taught
    artist, Beasley's Vaqueros is a record of vaquero life in South Texas
    that spans the centuries" —Provided by publisher.
Identifiers: lccn 2021044889 | isbn 9781625110572 (cloth) 9781625110701 (paper)
Subjects: LCSH: Beasley, Ricardo M., 1908-1994. | Drawing, American
— Texas, South—20th century. | Cowboys—Texas, South—Portraits. |
    Mexican American cowboys—Texas, South—Portraits. | Cowboys in
    art. | Ranch life in art. | Cowboys' writings, American—Texas, South. |
    Cowboys—Texas, South—Poetry. | Cowboys in literature. | LCGFT:
    Free verse. | Interviews.
Classification: LCC NC975.5.B42 T55 2022 | DDC 741.973—dc23/
    eng/20220301
LC record available at https://lccn.loc.gov/2021044889

CONTENTS

*Funding for this book was made possible by the generous support
of the Texas Historical Foundation and Mrs. Gloria López*

ILLUSTRATIONS

FOREWORD

RICARDO BEASLEY was not a familiar name to me until one day, more than two decades ago, when Andrés Tijerina took me on a whirlwind tour of Alice, Texas, the homeland of the "Duke of Duval County," George Parr. It was my first visit to the Duke's storied country, which I had read about in Robert Caro's multivolume biography of Lyndon B. Johnson and heard about first-hand from Edward Clark, a long-time Johnson associate and one of the attorneys who had fought to save the 1948 senate election for him. It was with some surprise that I learned that the artist whose work we had come to see was one of the original members of the Freedom Party, organized in 1952 to liberate the region from Parr's dictatorial rule, and that he credited the violent environment in Duval County as the source of the ferocity that is a hallmark of many of his drawings. And I was even more surprised to see the energy and intensity—the chaos— exploding from his work.

Ricardo Beasley was a South Texas vaquero, an instinctive but untutored folklorist and poet, the "artist of the brush country," and a lover of freedom. Born in 1908 near Kingsville, he descended from Spanish forebears who settled the region in the late eighteenth century and grew up in a unique culture—neither Mexican nor Anglo but Tejano. He developed his artistic enthusiasm from a set of encyclopedias that his mother purchased, and his earliest sketches were of friends of his vaquero father, cattle, and the brush country. As he grew older, he spent time at primitive cow camps and worked as a vaquero himself, savoring the courage and character of his comrades. In the few photographs of Beasley that are available, he appears to be short (like most vaqueros), bespectacled, and intense. As he closed out his own career on horseback, he realized that he was privy to a distinctive coterie of men who exhibited heroism every day in the face of extreme danger, whose natural adversaries were the wild steers and mustangs that had descended from the herds that the Spaniards had brought to the New World, and who had created the Tejano culture which nurtured him. That experience left him with a story to tell and gave him the confidence—the passion—to document their adventures.

Arrayed against the sinewy vaqueros were the Tejano longhorns, cunning, brawny beasts with lethal horns that pointed out and then up. Beasley pictured one such monster that killed three horses in a single encounter. The *mesteños* (mustangs) were equally fierce and combative, matched only by the determination of the vaqueros who, like their descendants, the latter-day cowboys, "could say the most in the least words." His hope was that today's children of the brush country, growing up as he did, would feel the "force" of the region through his drawings and understand the freedom that it represents.

Beasley has sometimes been compared to the most famous cowboy artist, Charles M. Russell of Montana, and the parallels are intriguing. Russell's ancestors had arrived in America in the seventeenth century, and he, too, was self-inspired and taught. As a youngster he got a job with a cattle outfit and began sketching and painting many of the incidents that he saw or heard about around the campfire, just as Beasley got much of his wisdom and subject matter in the brush country cow camps. Russell was equally gifted as a writer, and his colloquially eloquent, illustrated letters that are treasured by museums and collectors today foreshadowed Beasley's free verse descriptions of the daily contests of wit and will between men and beasts. Both Russell and Beasley preserved in their art the unique dramas they had heard of and witnessed.

Beasley died in 1994, and Andrés Tijerina became aware of his lifework shortly afterward, while he was researching the ranchos of South Texas. He included a few of Beasley's drawings in his book *Tejano Empire* and became captivated by them as he learned more about the man and his compulsion to record astonishing feats of the vaqueros in his forceful, action-filled pen-and-ink drawings: a man who as a child drew late into the early morning hours by the light of a kerosene lamp, a habit that he maintained all his life; who liked to quote Leonardo da Vinci and El Greco about the need for passion in one's work; and whose drawings seemed filled with wild horror or joy. Beasley would have published a book of his drawings if he had found a person he thought worthy of the task. Andrés Tijerina is that person. He shares the love of the South Texas brush country that disciplined and inspired Beasley, and, as a U.S. Air Force pilot and decorated veteran of one hundred missions during the Vietnam War, he experienced the same rush of adrenalin that Beasley's vaqueros felt as danger stalked. But, most important, Andrés shares the passion that animates Beasley's drawings and verses of the men and animals who were molded by the singular South Texas culture and, in turn, inspired the modern Texas ranching industry.

Ron Tyler
Retired Director
Amon Carter Museum of American Art

PREFACE

I FIRST DISCOVERED the art of Ricardo Beasley while conducting research for my book *Tejano Empire* in the early 1990s. In that book, I wanted to show that the true significance of the cattle kingdom and the colorful cattle drives of Texas longhorns was not just about the cows or even the cowboys. The six million longhorns provided the protein for the burgeoning industrial cities of the United States in the 1880s, and they changed the preferred meat of the American diet from pork to beef. But while other writers of the cattle kingdom saw only cows and cowboys, I focused on the ranching families who raised the six million cows. I found the thousands of Tejano extended families who stretched their *compadrazgo* (godparentage) networks from San Antonio down to Brownsville. I wanted to show how their strong family loyalty developed the community bonds and leadership that would ultimately give the United States its modern ranching industry, the League of United Latin American Citizens, the American G.I. Forum, Tejano music, countless Medal of Honor heroes, and the strong leadership to break down the discriminatory barriers to middle-class American life.

I probed their family structure, Spanish land grants, and solid homes made of limestone *sillar* blocks. And as I researched and interviewed for information on the *vaqueros*, the original American cowboys, I discovered Beasley. I interviewed old ranchers, the widows of old vaqueros, and archivists who knew about the land-grant families. And as I talked to these respondents, Beasley's name kept coming up. An old rancher would be describing an incident, and he would say, "Yeah, Ricardo Beasley did a drawing of that." Or a vaquero's widow would say, "Yes. My husband is the one in that drawing that Ricardo Beasley made of that incident." So, I began to ask, "Well, who is this Ricardo Beasley, anyway? And what drawings?" Then one day, I saw one.

I was interviewing a lawyer, Chuck Barrera of San Diego, Duval County, Texas, at the landmark Earl Abel's Restaurant on Broadway Street in San Antonio, where he had offered to meet me for the interview. Chuck was Beasley's close friend and lawyer. He showed me his original Beasley drawings. When I saw them, I immediately recognized the authenticity of the subjects. Indeed, in my research I had already learned many of the names of the great South Texas vaqueros such as Sam Smithwick, Teófilo Salinas, and Colón de la Garza. Now, I had Chuck Barrera pointing them out to me in the drawings. "That's Santiago Muñoz, saving Sam Smithwick's life. Beasley was there," Chuck said with a smile. "Here's the one of Juan Everett and Smithwick, roping the killer steer, 'El Mexicano.' He killed three horses in one day," he added proudly (drawing 10). It was apparent that Beasley represented an unsung Tejano artistic heritage in South Texas. This was not historical narrative. It was genealogy.

The drawings were snapshots of real-life men, in real time. One drawing captured the moment of a storied vaquero's death. And Beasley was there. On the back of one drawing, he wrote a description and named the vaquero. He recorded the name of the horse as an antagonist in the drama, "El Coyote." He described the horse's distinctive color, writing, "He was the color of a coyote, I knew him." As a vaquero, Beasley worked with the very figures in his drawings. And the drawings were dynamic, captivating. As I was finishing my manuscript for this current book, it struck me that many of Beasley's narratives were actually poems. As a historian, unfamiliar with free verse poetry, I had assumed that these were simply short narratives. But some guidance from a friend, Kurt Heinzelman, made it clear that most of Beasley's drawings are accompanied not only by his own recollections, but also by his own free verse poem about the incident. Beasley was an artist, a folklorist, and a poet.

The drawings captured the movement, the drama of roping a wild bull. They revealed the intensity in the bull's eyes and its deadly focus on the horse's neck—on the vaquero himself. As mid-action shots, the artist constructed four drawings, showing four separate phases of the drama—the goring of a horse, lifting the horse high in the air, driving the horse's body down in the ground, and a portrait of the monstrous beast with its head held high, standing over the remains of its lifeless victim. I never saw this in any of the popular Old West cowboy paintings. The dramatic action in his art is the South Texas answer to the romantic works of Eugène Delacroix and Théodore Géricault. The difference was that Beasley was a witness to many of these dramas, and these were not "Tiger Hunts" on a far-off continent. His drawings captured daily work, on Chuck Barrera's neighbors' ranches in Duval, Jim Wells, and Kleberg Counties in South Texas.

It was also starkly apparent that these were not the old Texas longhorns that we have all learned about in the cattle drives of the 1880s. These were the cattle that were left behind after the 1890s. They had interbred with the exotic stock like the Brahman bulls that were imported into South Texas by the King Ranch and other large corporate ranches. This was a completely isolated stock of cattle—the vaqueros did not call them cows. They called them *bestias* (beasts) or *novillos ladinos* (wild steers). They had been left out on the vast corporate ranchlands after the 1890s. By the 1920s when the ranches were being broken up into farms, the ranch homes had long been abandoned and the last of the open ranges were cleared of brush and derelict livestock. The cattle, thousands of them, had been left for decades, thirty miles away from the nearest small town. And the corporate ranchers did not live on the ranch like the original Tejano ranch families that they had replaced. Most of the livestock had lived their entire lives without seeing man or horse. And when they did, they attacked.

The descriptions come from the old vaqueros—not from veterinarians or historians—but from the men who roped the last of the wild steers, corraled them, and drove them to market. The descriptions are in Beasley's own narrative, his recounting of vaquero stories, and above all else his intricately detailed first-person drawings. In many cases, Ricardo Beasley recounts the tales of his vaquero friends. The vaquero accounts indicate that it was not only the size and bad temperament that these beasts inherited from the Brahmas. They were large and ferocious, to be sure, but the killer rack of horns made them truly dangerous. These were not like the wide, horizontal seven-foot spread of horns mounted over the mantle in the hunting cabins of proud Texas ranchers. These horns pointed forward. They were long, but not pointed outward. One old vaquero said they pointed forward and "curved up like a basket," as Beasley noted, to gore straight into the heart of the victim and then throw it overhead.

And facing these wild steers was a man on horseback—the vaquero. The word *vaquero* comes from the Spanish word *vaca* (cow), and it refers to the person who works the cow. Americans came to use the pidgin pronunciation of "buckaroo." Likewise, Americans used the modern word "Texan" as the equivalent to the Spanish word *Tejano* for the pioneers who founded Texas under the Spanish and Mexican flags. Beasley's vaqueros were the direct descendants of the original Spanish horsemen who first imported cattle to the New World. When Spanish explorers first came to Texas in the late 1600s, they brought their cattle with them as fresh beef on the hoof for their long expeditions. Captain General Alonso de León intentionally set cattle free on his expedition across Texas, and by the early 1700s, Spanish expeditions reported that wild cattle roamed the Texas prairies by the thousands. More importantly, they recorded that Tejano settlers were expert horsemen who would rope and round up the cattle and held them in makeshift pens made of brush. Thus, the Texas vaqueros established the fundamental aspects of North American cattle ranching: the roundup, the brand, the open range, and the cattle drive. The Texas longhorn and the cattle kingdom had their genesis in South Texas—not in Mexico or Spain. Spain did not have six million cattle, and Mexico did not have longhorns. The Texas vaquero was the origin of America's cowboy.

In all of the cowboy literature, the cowboy museums, and the western cowboy art, few if any artists painted the original cowboy—the Tejano—as a substantial figure in their portfolio. There are many paintings of Native Americans and the Anglo cowboys. A few artists included a painting or a bust of a Mexican vaquero. Indeed, Frederick Remington had various paintings of Mexican vaqueros, but these were disparate paintings in a wide variety of settings and states. No artist had a portfolio of vaqueros as a major figure in North American ranching history, or collections focusing on the Tejano—until Beasley.

Ricardo Moreno Beasley is the original vaquero artist; that is, a vaquero who was also an artist. The more than one hundred drawings he executed represent the only art portfolio of the original American cowboy. He was an on-the-ground witness to his subjects. He himself was a direct descendant of the original Spanish land-grant settlers in North America, and he worked as a vaquero. His drawings have a fidelity to the action informed by firsthand knowledge. He dedicated his entire life to his art with little regard for his own personal comfort. He chose to live for years in a cabin with his art. He shared his art, gave it, and lent it, but rarely sold it. He never presented it in a major commercial exhibition. Chuck Barrera was with him when he rejected a contract to publish his art in a book because he did not want his writing edited by a famous scholar who "never rode a horse." Beasley would be seen walking on the streets of San Diego or Alice, always with his art notebook under his arm. He never married, though he knew a woman's love, and loved his son. His art is scattered across South Texas, in the hands of individuals who proudly displayed his art. He died intestate. Upon his death, he maintained his refusal to capitalize on his art, which by then had become richly prized by connoisseurs and the San Diego ranching community. Like his beloved vaqueros, he never broke his principles.

As I learned more about Beasley's life and works, I identified several of his drawings that coincided with the vaqueros and incidents in a manuscript that I was finishing, which became the book *Tejano Empire*. At the last minute, I notified my publisher, Texas A&M University Press, that I would insist on including Beasley's drawings in my book. Mary Lenn Dixon, the editor-in-chief at Texas A&M University Press, relented but notified me that it would delay printing and release of the book. Her compositors incorporated eighteen of the Beasley drawings and used one of them for the front cover. I noticed later that my local Barnes & Noble bookstore always placed my *Tejano Empire* book on the shelf face-out to show the cover art while the other books were lined up on the shelf to show only the book spine. I then resolved that I would commit my next book to revealing Beasley himself to a world that had never known his gift.

Beasley's lifelong goal, as he wrote in "The Dream," was to show the world the dramatic life and noble heritage of the Texas vaquero. With limited schooling and absolutely no art instruction, this naive artist captured vaquero exploits on paper, wrote his narratives, and constantly related the knightly deeds of the vaqueros. My goal is to share Beasley with a fidelity not only to his goal, but to his purist standard of his writing, his spoken words, and his exclusive focus on the vaquero—not my interpretation, not Beasley himself, and with no appeal to a commercial market.

The problem I found sharing Beasley's legacy was that his portfolio was scattered across Texas in different hands, which mostly held only one or two drawings. His narratives were handwritten pages and divided among different notebooks, which ended up scattered among his cousin, his son, and others, with no one knowing what the other held. His poems were written as separate notes, some handwritten, others typed. I did not even recognize the poems for what they were until I shared one with a reknowned poetry professor, who immediately identified it as free verse poetry. I struggled for years trying to find a structure for a biography of him and his art. Then I had an epiphany.

I decided that my job would be to compile all those disparate drawings, notes, and poems. As soon as I organized them into themes of vaqueros, mesteños, wild steers, and landscape, I began to realize that the different media were interrelated. Each drawing had a short narrative that described the action, the vaqueros, and the background story. And many of these drawing-narrative pairs had a matching poem. My job was to coordinate the different media and organize the themes as chapters, and as a historian, to write the historical context for each chapter. The crowning discovery was a recorded interview in which Beasley described his work in his South Texas vaquero Spanish, which employs many terms that are archaic or peculiar to South Texas. After transcribing the recorded interview, I realized that it too had descriptions that corresponded to the drawings, written narratives, and poems. With my transcript of the recording, there were now four media—five, counting my English translation of the transcript.

This book is the compilation of Beasley's work in different media. His art reveals his subject as well as his own artistic technique. Some of his narrative writing was preserved in a notebook copied and compiled by his cousin, the late Manuel Peña of Alice, Texas. His son, Ricardo "Rick" García, also provided me with many poems, writings, and narratives that he had in his possession. The notebook contains Beasley's poems along with the pages of a narrative. The narrative describes the art's subject and the poetry captures the spirit. Finally, a major portion of this book consists of my transcript of the one-and-a-half-hour tape recording of an interview. In 1979, a San Diego banker and realtor, Tommy Molina, had the vision to preserve Beasley's own verbal narrative on tape, in his own words. Tommy himself is a rancher and direct descendant of the founding families of Texas under Spain. His grandfather is the Tomás Molina mentioned in Beasley's narrative. In the interview, Tommy systematically prompts Beasley to explain one drawing after another and to narrate the details in the incident as well as his artistic technique. Beasley responds with a heartfelt enthusiasm, unleashing an uninhibited laughter and expressing his own amazement at the vaquero talent. He articulated this narrative in a South Texas ranching lexicon that has dropped from modern usage, along with the vaquero ranches, lifestyle, and skills. Many of Beasley's terms and expressions reflect an archaic Spanish that was brought directly from Spain to South Texas in the 1750s, many of which are peculiar to the vaquero practices on the remote ranges. Like the vaquero accoutrements, many of the terms derive from the ancient Arabic school of horsemanship brought from Spain. It is difficult to find anyone alive today who would know or use these terms in modern America.

Tommy Molina preserved Beasley's conversation in Spanish. I transcribed the entire recording and translated it into English. This book then, consists largely of the words of Ricardo Moreno Beasley, including the English transcript, the Spanish transcript, his narrative, his poems, and his art. My only contribution was the organization of the art and the editing of Beasley's text. I corrected a few random words, spellings, and phrases only when necessary to make them understandable to the modern reader. One message I got very clearly is that Ricardo Beasley did not want anyone to write his book who "never rode a horse." I did ride horses and work cattle as a youngster in West Texas ranches, but I still do not presume to tell his story. I published this book with the goal of introducing the genuine Ricardo M. Beasley. And Ricardo M. Beasley turns the spotlight on the genuine vaqueros.

I've written a historical context to introduce the theme of each chapter. I signal the shift from one medium to the other with notations like "From the notebook of R. M. Beasley," "Transcript: R. M. Beasley Interview," to guide the reader from the narrative to its corresponding drawing and poem or transcript passage. My hope is that the reader can perceive the chapter theme streaming through the distinct media and ultimately grasp Beasley's vision.

There are two distinctive aspects to Ricardo Beasley's art that I will interpret for the reader, even though the art presents them in stark, even shocking reality. One is the ferocity of the animals and the other is the valor or the vaqueros. The Texas histories by Walter Prescott Webb and J. Frank Dobie speak of the Texas longhorns and admit that these were rugged animals. I even had the great honor of knowing the late Enrique Guerra of Linn, Texas, who had a virtual museum in his home of Tejano ranching artifacts based on his family's Spanish land-grant heritage. Enrique is the individual who rebuilt the DNA of the Texas longhorn stock after the original stock had all been driven up the cattle trails to be eaten by a hungry American population. His longhorns are certified, but even his bulls stood passively as we drove past them. Not until I studied Beasley's drawings did I know about the hybrid cattle that replaced the longhorns on the corporate ranches. Beasley offers a close-up view of the breed that emerged on those remote ranges. His narrative depicts a ferocious king of the open range, the *novillo ladino* (wild steer).

To truly appreciate the valor of the South Texas vaquero, Americans will have to unlearn certain stereotypes of Tejanos and Mexican Americans in general. For example, many Americans seem to perceive Mexicans as malicious or brutal, with the image of Mexican general Antonio de Santa Anna or a crude, unsophicated bandito as the archetype. More positively, but still incomplete and paternalistic, they might find it easy to accept an image of the Mexican as an innocuous, subservient figure—someone fit for drudgery or to be underlings, at best. The Mexican vaqueros of the King Ranch, for example, are often described by writers and scholars as skilled but loyal workers. Beasley's vaqueros stand in stark contrast to these passive figures whose highest attribute is that they are simple and loyal.

Beasley's vaqueros possess that quality that Ernest Hemingway admired when he stated, "Auto racing, bull fighting, and mountain climbing are the only real sports . . . all others are games." These vaqueros possess the quality that I observed in my fellow U.S. Air Force pilots in combat during the Vietnam War. These pilots were the highly trained, college-educated officers who flew the most advanced supersonic aircraft in the world. After flying combat missions, we gathered in the stag bar, sang barroom songs like Neil Diamond's "Cracklin' Rosie," and hoisted our beer mugs—sometimes in memory of a buddy who had been shot down that day. These men were the best the United States had to offer. They were young, educated, and they gave much more than just loyalty. They faced death and executed the mission. But afterward, they laughed. They told stories. The same stories they'd all heard before and gradually grew together to form comradery, to share feelings, and to steel their resolve against fate. Beasley's vaqueros, like Hemingway's bullfighters, racecar drivers, mountain climbers, and American fighter pilots, were fiercely competitive. They daily stood in the teeth of danger and lived by virtue alone of their superior skill, their sensory acuity, and a spiritual integrity. As Beasley's mother told him, their life "hangs on a thread" and even "the best get killed."

Most Americans did not know these vaqueros, but Chuck Barrera knew them. Beasley knew them, too, and saw them live and die in their daily challenge against nature. Though they might be isolated in remote gullies deep in the chaparral, they affected their broader Tejano community. Beasley depicted the vaqueros in his unique portfolio, but he only reflected the respect that vaqueros enjoyed within the broader Tejano community around Duval County and South Texas.

Andrés Tijerina
Austin, Texas
2018

ACKNOWLEDGMENTS

I MUST ACKNOWLEDGE that I did not have the insight to see the rich, living heritage of Tejano South Texas until I married into Juanita García's family in San Diego. We were married in the century-old church that is depicted in Beasley's narrative of the cattle stampede through the middle of the town plaza in front of the St. Francis de Paula Catholic Church (drawing 13). Shortly after that, my father-in-law told me about the stampede. Through the years, when we went to visit Juanita's family on a ranch north of San Diego, I sat around the campfire and listened to him relate the stories and the legends about Don Sam Smithwick and the original ranchers of Duval County. My mother-in-law told us about the Guerra and Solís sides of her family, who came with the original Spanish ranch settlers in the 1750s, and who later intermarried with the Smithwicks and the Dixes. To my wife's family, I owe my first introduction to this rich Texas history.

I also have to give thanks to Dr. Ronnie C. Tyler, the former director of the Amon Carter Museum of American Art who accompanied me on a trip to San Diego when he was executive director of the Texas State Historical Association (TSHA) to evaluate Beasley's portfolio. As a nationally recognized expert on western art, Ron confirmed the value of Beasley's art, and gave me the first endorsement to create this book for the TSHA. Publishing with TSHA is the highest honor for any book on the history of Texas, and I am honored to be associated as a TSHA member, a Fellow, and an author. Ron also put me in contact with Scott Barker of the Dow Art Galleries in Fort Worth who found a *Fort Worth Star-Telegram* article from February 15, 1948, and a master's thesis confirming that Beasley did have at least one major exhibit for the Fort Worth Art Association in 1948. The article confirmed my opinion that Beasley's works are equal to, and occasionally even excel, those of Remington and Russell.

The art in this book would not be available if it had not been for Beasley's son, Ricardo García. He opened the entire portfolio of original art to me for this book. More importantly, he shared with me the discipline to be true to Beasley's principles. Tommy Molina not only gave me a copy of his taped interview with Beasley, but he helped me identify people to interview, and Tommy took me on the old Beasley ranch. Walking gingerly on the pulpy wooden floor, I found Beasley's old bulletin board with a note that said "Call Elena" and art magazines from the 1960s. I realized what Beasley meant in his narrative when he said that the *chaparro negro* (black brush) was dense and impenetrable. I could only imagine a vaquero riding full speed into this tangled thorn brush to rope a 2,000-pound wild steer. Tommy gave me a list of other contacts in San Diego who hold Beasley originals.

I was guided on a personal tour of San Diego and Alice by Homero Vera, who is also descendant of the original Spanish ranch families. Homero granted his good offices for me to enter the home of a vaquero's widow to interview her. Homero is now the manager of the Kenedy Ranch, one of the largest corporate ranches beside the King Ranch. His magazine, *El Meste*ño, provided a good background on the life of Beasley and other ranchers of South Texas. Homero is in many ways the gateway to the Tejano history of South Texas. Chuck Barrera gave freely of his valuable lawyer time to interview him on Beasley's personal life. Chuck shared numerous gems of Beasley's thoughts and experiences that informed my research and writing.

Other people who played crucial roles include Lupe Martínez and the late Manuel Peña, both of Alice, Texas, who gave me copies of Beasley's art pieces and writings. The late Alice, Texas, attorney Héctor López first let me see the bulk of Beasley's portfolio, which I indexed to grasp the size and variety of Beasley's work. Frank and Jeanne Egbert Wright shared copies of Beasley's original art and his letters. I met them through their daughter Elizabeth, who like others e-mailed me to say she had heard I was doing a book on Beasley. For professional advice, I have to thank my colleagues at the Texas Institute of Letters and its former presidents of the Texas Institute of Letters, W. K. "Kip" Stratton and Carolyn Osborne. On long trips to the TIL council meetings, they listened as I told them the highlights of my manuscript. Both of them kept reminding me that they were excited to see the finished product. Kip read an excerpt of my early manuscript, and he made valuable suggestions on organization, reminding me that the book would be an important addition to literature and art. Their support reinforced my drive to complete the work. These friends are all proud Texans who share my commitment to telling the world about Beasley and his vaqueros.

I want to express my genuine gratitude to Marise McDermott, president and CEO of the Witte Art Museum of San Antonio and to Bruce Shackelford, the chief curator at the Witte, for giving me their endorsement and full support in my initial discussions with the TSHA for this book. Marise and Bruce lent invaluable credibility to my Beasley art book proposal, reinforcing that with discussions of an art exhibit.

Finally, to the people of San Diego and South Texas who share my deep appreciation of Beasley's work and who proudly offered their original drawings and support, I write this book for them. Beasley is their son. He is telling their story, and their trust is a cherished responsibility for me. It took me many years to compile the materials, and to publish this book, but I now know why I could never get myself to write it. I finally opened my eyes and my heart to realize that it is not about me. It is not about Beasley. The book is about the vaqueros, and nothing and no one can speak better than their spirit coming through pure art. Let it speak for them.

Editor's Note

MOST OF THE TEXT is the writing and words of Ricardo M. Beasley except for the preface, the introduction, and the editor's notes for each chapter. I also inserted a few brief explanatory notes in the text with an asterisk (*). Some of his narrative is from his own notebook and some of it is my own transcript of his tape-recorded interview mostly in Spanish by Tommy Molina. I include the transcript of his interview along with my own English translation in the bordered boxes to accompany the art and text. I have edited only a few of his words and added the subtitles to guide the reader through the different formats of text. His misspellings and grammatical errors have been left intact. Otherwise, the entire book consists exclusively of Beasley's words in the following formats:

Narrative: Beasley's text labeled as "From the notebook of R. M. Beasley."

Poems: Beasley's original poems, some handwritten, related to the drawings.

Art: Beasley's original drawings of incidents on the ranches, most of which he witnessed.

Transcripts (English and Spanish): My transcription of Beasley's one-and-a-half-hour recorded interview with Tommy Molina are labeled herein as "Transcript: R. M. Beasley Interview."

Introduction

AMERICANS KNOW the the colorful Texas cowboy, the popular Old West ranching magazines, and the pulp fiction cowboys of Charlie Siringo and Andy Adams, but the Texas vaqueros have been left out of the dominant narrative of the American West. Countless Americans have seen eight-second bull rides on animals that have to be coaxed by rodeo clowns to stay in the arena. But few people ever saw a vaquero on foot bulldog a steer by hand. Ricardo M. Beasley sought to address this debt of recognition and the unsung contribution of Texas vaqueros to the national ranching heritage. He never revealed the slightest resentment or bitterness, but only sought to show the true horseman of Texas. That was his life commitment.

Ricardo Moreno Beasley was born in 1908 near Kingsville, Texas. He lived all his life on a South Texas ranch and in San Diego, Texas. San Diego is a small ranch town founded in the early nineteenth century by early Spanish ranching families. His great-grandfather, Santos Moreno, received the La Trinidad land grant from the king of Spain and married the heiress of another Spanish land grant, La Mesteña. His paternal grandfather, Abraham Beasley, had come to South Texas during the U.S–Mexico War in the 1840s and stayed. Abraham married and had a son, Richard J. Beasley, who married Concepción Moreno, and had five children including their son, Ricardo Moreno Beasley, who would become the geat vaquero artist. Beasley was brought up on

La Trinidad Spanish land grant and identified as a Mexican American. His generation in San Diego and South Texas was interspersed with Anglo American surnames like Everett, Watson, Smithwick, Dix, Stillman, and Rogers, as Americans intermarried into Tejano families and identified as Mexican American.

This stock of Texans remained oblivious to any of the ethnic labels that categorize other modern Americans. Beasley said, "They had no connection with either the Mexican or Anglo culture." They are descendants of the founding Spanish families who have never urbanized, physically or mentally. Their identity is not defined by the Alamo or the civil rights movement. When the time came to name their high school football team, they thought nothing of calling themselves the San Diego Vaqueros. They could not imagine calling their students lions, eagles, or rebels. These are the original settler families of Texas, proud of their Spanish and Mexican heritage, and proven in their military service to Texas and the nation.

This same Texas community produced American soldiers with more than their share of combat decorations and Medals of Honor. These are the same vaqueros who went south across the Mexican border in 1862 to volunteer as cavalry to help the Mexican army defeat the French at the famous Battle of Puebla, which is now commerated on Cinco de Mayo. Their cavalry leader was Porfirio Zamora, who married Ricardo Moreno Beasley's great-aunt, Gabina Moreno. The same Porfirio Zamora whose daughter, Elena Zamora O'Shea, wrote the first book published by a Mexican American in Texas, *El Mesquite*, which is about Tejano vaqueros and ranchers.[1] Elena, the first vaquero author, was first cousin of Ricardo M. Beasley, the first vaquero artist.

Beasley was proud of his Zamora line, but he also honored his Moreno heritage. One of his drawings depicted the family legend of his great grandfather, Santos Moreno. According to the legend, Santos had vowed to build a chapel for the ranch on La Trinidad Spanish land grant but died without finishing the chapel which fell into ruins. In his poem "Rest in Peace: Freedom Lives," Beasley spoke of Jabalín, the mesteño that roamed the ranch, and was last seen standing in front of the church ruins. Beasley's rendering (drawing 27) depicts a majestic black mesteño stallion, standing in front of the ruins and a tombstone with the inscription "Santos Moreno QDEP" (QDEP is the Spanish equivalent of RIP). His poem states that the mesteño "had to be where he stood," in tribute as "a last goodbye" or, in Spanish, "El Último Adios." The mesteño's tribute came as "an orphan . . . or that of a child abandoned at the door of a convent." The poem speaks of Moreno's legacy to his country and his family.

The land was one of the strongest identifiers in Beasley's works. San Diego is in the geographic center of the "Cattle Diamond" of South Texas. It was founded by the pioneering Spanish families recruited by José de Escandon, the Count of Sierra Gorda, in the 1750s. The direct descendants of those original families are one of the largest populations of Americans to remain in their ancestral homeland. In his poems and his narrative, Beasley articulates the unique language of heritage, landscape, and animal life of the South Texas brushlands. These are the people who bred the unique Texas longhorn and the distinctive Texas mustang (mesteño). His art illustrates this dusty, rocky ranch country blanketed by dense brush, called black brush or chaparro negro. The *potreros,* or horse pastures, frame his art, and he sets the vaquero action richly in the terrain and the brush. Like the classic Texas writers Walter Prescott Webb and J. Frank Dobie, Beasley used the environment to inform his characters.

Dobie is deeply embedded into Beasley's vaqueros, not only because they worked on Dobie's Ranch, but because Dobie grew out of this vaquero culture in Duval County and Alice, Texas. Indeed, Dobie's school teacher in Alice, Texas, was Beasley's cousin, Elena Zamora O'Shea whose own book, *El Mesquite,* is based on the mesquite trees and landscape of Duval County. In his description of the famous King Ranch artist Tom Lea, J. Frank Dobie could easily be describing Beasley's art. Dobie states, "These pictures are the expression of a man who understands a vast land peculiar unto itself, a land with a culture of its own that he burns to reveal and make the dwellers in that land understand as their own inheritance."[2] Dobie could well describe his own folklore of the vaquero and the longhorns as being strongly informed by his own upbringing in Alice and Duval County under his teacher, who opened her personal letters to him as "Miss Elena," and closed them with "*Tu maestra vieja.*"[3]

Although he was proud of his Tejano heritage and lived as a ranch owner and vaquero, Beasley was hesitant to call himself a vaquero. He set a high bar for that profession and believed that a vaquero was defined by definite skills, feats, and character. He learned to draw as a youth and perfected his technique of pen and ink on white tablet paper, which he used all his life. He was a prolific artist starting from the 1920s through the 1960s, but he suffered a back injury in 1969 that severely limited his drawings in number and quality. He died in 1994 and was buried on his ranch, unpublished and generally unknown as a professional artist.

Ricardo Beasley was active in South Texas politics in the years after World War II. By 1950 he became deeply committed to freeing Duval County of a powerful political boss, George Parr. Parr's political machine dominated county politics for decades, and was notorious for using voter fraud, intimidation, embezzlement, and murder to perpetuate its power. In 1952, Beasley helped organize a political faction, the Freedom Party, to challenge Parr's power. Beasley's personal notebook contains a letter that he wrote to request voter protection by the Texas Rangers during the election as well as a photo of the Rangers camped out in Duval County in response to his request. His notebook also had a photograph of the Freedom Party's founding members, including Beasley and his father.

Parr targeted Beasley after he hosted the meeting. Beasley's lawyer, Chuck Barrera, stated in an interview that the sheriff put out a contract on Beasley's life, which led to at least one failed attempt on his life.[4] His poems and narrative contain diatribes against Parr's political machine, which include the following lines,

A corrupted press of a Gobernment [*sic*] destroys
his own people,
It keeps them intoxicated
in a false belief of superiority
And a great understimation [*sic*] of those they oppose
Because they can not dominate
Both shots have a destructive reaction.
[dated] 3/15/1962

Beasley backed away from political and even social involvement after the 1960s. Leaving the ranch, he dedicated his life to his drawings, which he shared with his community and local newspapers, many of which graced the walls of public places and private homes.

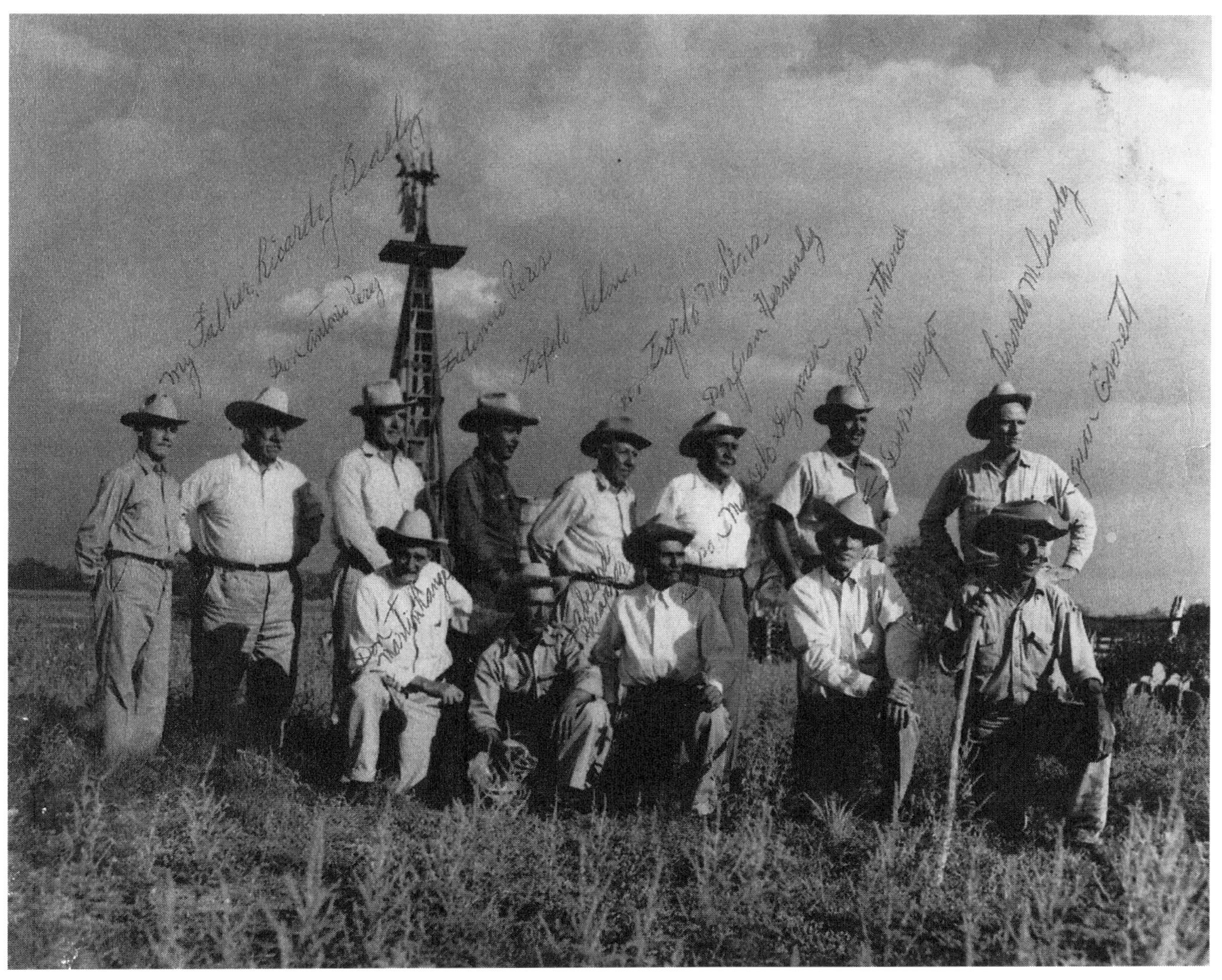

PHOTO 1

On the back of the photo, Beasley wrote, "Ricardo M. Beasley, Octover [*sic*] 13, 1951, Duval County, Aguilia Ranch. This picture was taken at the place when I honored some old vaqueros with a supper. Those present were my Father, Ricardo J. Beasley, Don Antonio Peres, Sr., Fidencio Peres, Teofilo Salinas, Don Teofilo Molina, Don Juan Hernandes, Joe Smithwick, Don Martin Rangel, Abelardo Guajardo, Don Marcelo Guzman, Doss Seago, Juan Everett, and myself." Ricardo M. Beasley is in fact, in the photo at the top right. He added, "From this meeting was the first time George Parr smelled the fire of the Freedom Party coming, the next day he called Joe Smithwick to find out if it was a political meeting. We started the Freedom Party March 1952. [signed] Ricardo Moreno Beasley, May 23, 1970." *Courtesy of Tommy Molina.*

One district judge stated in an interview that he once spoke to Beasley about his land. He said that Beasley told him, "Well, let the kids have it. They need it more than I do." [5]

Although Ricardo Beasley did not obtain first-class schooling, he was highly articulate in English and in Spanish, and he read books and art magazines prolifically. In an interview with Tommy Molina, he stated, "I went to school for the first time in English when I was fourteen years old" in Ben Bolt, a ranch village south of Alice, Texas, near his home. He wrote prodigiously despite his limited formal education. He kept a collection of his poems and his writings in a notebook and "Talley" books. In his own handwriting, he wrote the following poem in his notebook, revealing his dedication to informal knowledge gained from vaqueros:

Yo naci para los Vaqueros,
Cuando uno se pierde en
Un amor. No le echas
Menos al mundo.
Ricardo Moreno Beasley

Ricardo Moreno Beasley produced art. He secluded himself in a cabin deep in the ranch country where he worked for years, alone much of the time. He rejected social and financial encumbrances such as marriage and gainful employment—anything that would be an impediment to his art. He described his dedication to his art in recalling what he once told a girlfriend. "'Look,' I say, 'between you and my art, my art comes first. . . Like the Indian over in South Dakota,' I told her, 'before getting married,' he told his woman, 'Get this straight. The mountain comes first. The children come next. And then you come next.'"

"The mountain comes first." Beasley depicted the vaqueros like Hemingway's mountain climbers, racecar drivers, and bullfighters. They were dedicated to something, both physical and intangible, they valued above themselves and their lives. And though he said that he was not good enough to be a true vaquero, through his art and writing, Ricardo M. Beasley created, promulgated, and preserved the Texas vaquero life. He not only risked his life out on the remote ranges, working and drawing with his subjects, but he suffered a life-changing injury. He dedicated his life, his fortune, and his name to preserving this hidden history. Like the jet pilot who slipped the bonds of earth, flew skyward, and touched the face of God in John Gillespie Magee Jr.'s poem "High Flight," so did Ricardo M. Beasley touch the spirit of the vaquero, and preserve it for posterity.[6]

Ricardo Moreno Beasley is the only artist with a portfolio dedicated to the vaquero. Other well-known or famous artists have made a painting in their careers that depict a vaquero, but most portray the vaquero as an anonymous figure or only as a bust, and none depict the Texas vaquero. Beasley's portfolio includes more than one hundred drawings of the Texas vaquero and many others that portray the animal life on the South Texas brushlands. Likewise, there are many books of cowboy art or western art, but most have no Hispanic subjects or only a single painting of a vaquero.

None of the best-known books on western or cowboy art is dedicated to Texas vaqueros. The Texas Cowboy Artists Association published *The Texas Cowboy,* which lists the works of fifteen artists, but features no Hispanic artists or subjects. *The Cowboy in American Prints*, edited by John Meigs, identifies only one or two peripheral figures in all the drawings it includes as Mexican vaqueros. The book considered one of the best historical books on the history and methodology of cowboy art and the best survey of cowboy artists is *The Cowboy in Art* by Ed Ainsworth. This book reviews the cowboy's horse, cattle, and accoutrements, as well as the vaquero legacies of the corral, the saddle, the roundup, and the mustang, but it has no mention of the art of the vaquero. Even Mexican American scholar Américo Paredes in his article "El cowboy norteamericano en el folklore y la literatura" from a scholarly journal, *Cuaderno del Instituto Nacional de Antropología*, makes no mention of the Texas vaqueros and no mention of Ricardo M. Beasley.[7]

Many other artists in the last century have painted ranching livestock or scenes of the American cowboy or western themes, but none has painted the Texas vaquero. Frank Reaugh is known as the Texas longhorn artist. Indeed, J. Evetts Haley once stated that Reaugh was the "first great painter to see the artistic possibilities of the lanky Longhorn steer."[8] Another famous artist of the Spanish and Mexican ranching frontier is José Cisneros. Cisneros painted formal renditions of colonial soldiers and other figures in the northern Mexican border provinces. He was perhaps the best known of Mexican American artists, but there is no indication in his artwork or the books written about him that he knew of Beasley or the Texas vaqueros of the early twentieth century.[9]

Other artists with works on cowboys or western ranching themes include W. Herbert ("Buck") Dunton, Edward Borein, and the best known, Frederic Remington, Charles M. Russell, and Ernesto Icaza. Borein did some works on vaqueros, but they focused only on the vaqueros in California. Likewise, Dunton's significant body of work was devoted to vaqueros from the Babicora Ranch in Chihuahua, Mexico. And while Remington depicted Mexican and Texas vaqueros in a significant body of work, he and Dunton depicted vaqueros in Texas as "Mexican." Indeed, most of these artists around the turn of the twentieth century neglected the largest cadre of vaqueros—the Tejanos. Among the exceptions were H. D. Bugbee and Joe De Yong, who had a large body of work devoted to Hispanic vaqueros and gauchos.[10]

Beasley enjoyed only one major exhibition of his art, in Fort Worth. The exhibition inspired high praise. A writer for the *Fort Worth Star-Telegram* wrote that "his drawings were the sensation" of the exhibit, "even excelling, in some instances that of Remington and Russell." He never studied art in school, at an art studio, or with another established artist. One local interview respondent stated that he thought Beasley had gone to New Mexico to talk to painter Peter Hurd, but that was never confirmed. The only frequent source of publicity for Beasley was the local press. Numerous articles appeared in the regional newspapers, describing his unique style and subject. Outside of Beasley's home region, one article on the artist appeared in the *Houston Chronicle*, and another was supposed to be printed in a San Antonio newspaper, but the draft died in his notebook because he never approved it for publication.[11]

Ricardo M. Beasley had a unique subject in his vaqueros, but he also had a distinctive method. His perferred medium was black pen on white tablet paper except for a short period in the 1930s when he used a sepia paper. He called his method "Pen and Wash." In his taped interview with Tommy Molina, he said he always used black ink. He said, "The same, the same ink, I put it in little cups. I use a tile that keeps it fresh, compartments, and I add water and I go on making a kind of a clearer tone, clearer, and more clear. Understand? After the sketch is drawn by pen, I give it a very light wash. Understand? That gives the drawing a different look. It gives it a look of etching. It looks bolder." Asked if he ever tried any other method, he replied tersley, "No. It's the pen and paper." He said he secluded himself in a small shack on the ranch stating, "That is my camp in Duval. That is where I learned to draw in action. I was there in that little ranch eleven years." In a 1937 drawing entitled *The Old Home* (drawing 1), he drew the original Beasley house on the family ranch near Ben Bolt, after which a modern home was built.

He further explained the importance of the house when he described *Fallen Cow* (drawing 15). He said, "This is where I learned my drawing in action. I found the fallen cow, the one I showed you a while ago. I learned to give it movement. When I first arrived here to these fields,

I wasn't able to sketch in action. But there in that little house was where I achieved it all. I lived it all. It was my camp. It was everything. The cows would come and hunker down next to the house, the horses would come in at dawn and bang at the door . . . I wanted to live life now, with everything wandering around me. I had a camp box inside, some chairs, a water barrel. We had anything anybody [would want], a camp. But . . . everything simple. Those were the good old days. Juan Everett lived there with me for three years. In other words, that little house made me, well, it extended me quite a bit. There was something, just the way I wanted to live, it was just the right place. Even though there was a big house right there nearby, the little old lady landowner would tell me, 'Use the big house.' I said, 'No, I feel very comfortable right here.'"

As *Fallen Cow* shows, one of the other distinctive characteristics of Beasley's drawings is the movement, the action. He indicated that it was another ranching incident that inspired him to develop his technique of depicting action and movement. He said, "In 1946 was when I started sketching in action. I say with action. And it started developing

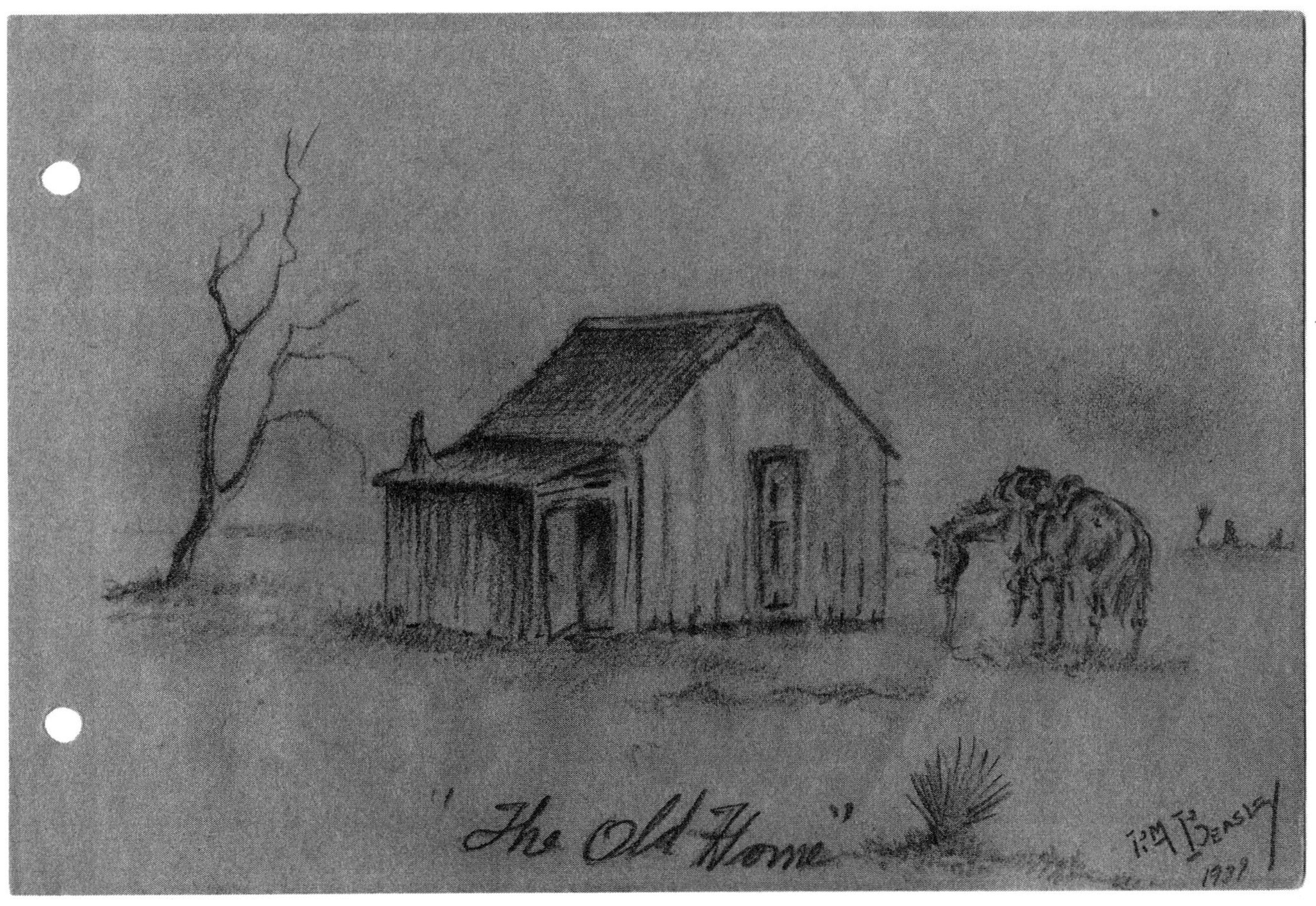

DRAWING 1
The Old Home. © R.M. Beasley Legacy LLC. Reprinted with permission. *Courtesy of Jeanne Egbert Wright.*

very quickly. Already by about, around, by forty-seven, forty-eight, well, in that one where Santiago Muñoz is with the steer, where he saved Sam, I did that one in those years. It was in those years that I acquired that technique."

When asked about the most critical aspect of his art, Beasley answered, "Danger. Always danger." He added that he learned to employ the juxtaposition of contrasting affective images, citing his drawing of the "San Diego Stampede" in front of the century-old Catholic church set in the town plaza (drawing 13). "You can see that church in the background . . . That's what appealed to me about sketching them. That with that church, it looks more violent. It gives it . . . it looks a brutal sight for something like that to occur at a church. Because for me, it's a theme, a terrible theme. Because without the church there, it wouldn't look as violent as it looks."

While Beasley always stressed the human figures in his dynamic art, he specified that his hallmark was the supremacy, the ultimate dominance, of nature over man. "I always give the wild animals the upper hand because they are the only ones in the right," he said. His drawings reveal that consistent theme as the threatening steers stand ominously over their victim, whether man or horse.

Finally, Beasley's art and discourse seem to reveal a consistent pattern of anthropomorphosis, that is, speaking of his steers as "he" with human values and feelings. Throughout his poetry and his recorded conversation, Beasley regularly refers to the steers as "he." While this may be attributable to some extent to the fact that the Spanish "el" can be used as "he" or as "it," Beasley also used "he" in writing and in his English dialogue to refer to animals. In reference to a wild steer, for example, he states, "Only after he senses no presence of vaqueros for several days does he move toward some water hole." An even stronger expression of anthropomorphosis is revealed in his description of the animals as thinking with reason and using human cultural practices. He states, for example, "Some wild steers showed a sense of brotherhood among themselves." Many of Beasley's memoirs are informed by his vaquero friends like Teófilo Salinas and Juan Everett; therefore, this practice may well have been the norm among the vaqueros. In the story of "the Invisible Mule," for example, Juan Everett describes the steers as strolling along and stopping to chat with "friends" they encounter on the trail. He says, "they entertain themselves with other steer friends who are going toward the water, like someone who stops to chat with their friend that they meet on the same road."

While these examples of anthropomorphosis may be interpreted as the provincial culture of a quaint rural worker in remote, isolated conditions, they tend to enhance Beasley's expression of respect, heritage, and commitment. These are the qualities of Beasley's vaqueros. The actions of an "Invisible Mule" may have as much to convey for the South Texas poet as the rodent in Robert Burns's "To A Mouse." Beasley relates a certain universality in Juan Everett's departing thought that life is a "very vast range" and the trail around horse pastures can be "so very long." Likewise the simple line drawings with a pen on a sheet of white paper may in fact represent the rich spirit of a silent heritage.

NOTES

1. Andrés Tijerina, historical introduction to *El Mesquite: A Novel* by Elena Zamora O'Shea (College Station: Texas A&M University Press, 2000).

2. Kathleen G. Hjerter (comp.), *The Art of Tom Lea* (College Station: Texas A&M University Press, 1989), x.

3. The Dobie Papers at UT Austin contain several letters in which Elena Zamora O'Shea closes as "Tu maestra" or "Tu Madrina" (godmother). J. Frank Dobie Papers, Manuscript Collection MS-1176, Subseries C. Recipient. Harry Ransom Center, The University of Texas at Austin.

4. Chuck Barrera to Andrés Tijerina, Jan. 19, 1997 (interview); Connie Lu Beasley to Andrés Tijerina, Aug. 15, 2002 (interview).

5. Guadalupe T. Martínez to Andrés Tijerina, Dec. 20, 1999 (interview); Adan Galván to Andrés Tijerina, Mar. 15, 2002 (interview).

6. Roger Cole, *High Flight: The Life and Poetry of Pilot Officer John Gillespie Magee* (Stotfold, UK: Fighting High Publishing, 2014).

7. Américo Paredes. "El cowboy norteamericano en el folklore y la literatura." *Cuaderno del Instituto Nacional de Antropología* 4 (1963): 227–240.

8. J. Evetts. Haley, *F. Reaugh: Man and Artist* (El Paso: Carl Hertzog, 1960).

9. Félix D. Almaráz Jr., *Borderlands: The Heritage of the Lower Rio Grande through the Art of José Cisneros* (Edinburg, Tex.: Hidalgo County Historical Museum, 1998).

10. Harold McCracken, *The Frederic Remington Book: A Pictorial History of the West* (Garden City, N.Y.: Doubleday & Company, Inc. 1966); Douglas Allen, *Frederic Remington and the Spanish-American War* (New York: Crown, 1971); Louis Chapin, *Great Masterpieces by Frederic Remington* (New York: Crown Publishers, 1979); Aldrich Lanning (ed.), *The Western Art of Charles M. Russell* (New York: Ballantine Books, 1975); Edward S. Spaulding (ed.), *Edward Borein, Etchings of the West* (Santa Barbara, Calif.: Schauer Printing, 1950); Harold Davidson, *Edward Borein, Cowboy Artist: The Life and Works of John Edward Borein, 1872–1945* (Garden City, N.Y: Doubleday, 1974); H. D. Bugbee, *Branding with Pen and Ink* (Canyon, Tex.: Panhandle-Plains Historical Museum, 1980); Frances Battaile Fisk, *A History of Texas Artists and Sculptors* (Abilene, Tex.: Frances Battaile Fisk, 1928); Marjorie Von Rosenberg, *Artists Who Painted Texas* (Austin: Eakin Press, 1997).

11. "Special Alice Youth Rodeo Supplement," *Alice News*, June 2, 1966; "Ricardo Beasley Art Depicts the Drama Action of the Cowboy," *Alice Echo-News*, April 28, 1966; "Better Ranches & Farms," *Corpus Christi Caller Supplement*, Feb. 25, 1959; *Houston Chronicle*, "Rotogravure Magazine," May 23, 1954; *Fort Worth Star-Telegram*, Feb. 15, 1948.

Prelude

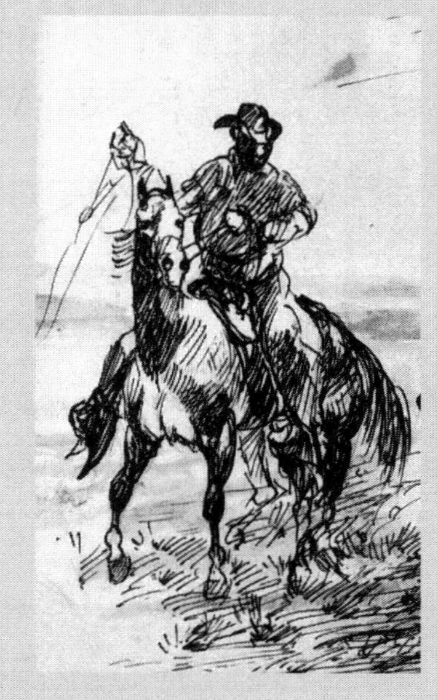

Editor's Note

Ricardo M. Beasley was in his late fifties when he wrote these candid but heartfelt words from his dream. Clearly, however, the seeds of this dream were planted when he was sixteen. It was at that age that he saw his destiny to reveal the majesty of the South Texas vaquero to the modern, urban youth who might otherwise never have known the proud heritage of the vaquero. He envisioned himself writing "a book to help children." It was only later that his father suggested that he use art to enhance his meaning.

Beasley wrote his narratives in a variety of styles. He augmented his message with art, but his narrative was so formal that much of it evolved into free-verse poetry. He did not refer to his poems as poetry. In fact, he wrote that he would narrate "in a way of song." But he did write in stylized phrases formatted as stanzas.

His narrative was as philosophical as it was formal. Even his narrative lines were phrased to emphasize descriptive terms or qualities of his subject. He framed heroic figures in a romantic lexicon. Beasley intensively crafted his art, his poems, and his formal narrative to sculpt a portrait of the vaquero. His words were not superfluous. They painted his subject—the vaquero—as intricately as his pen and ink drawings.

Rough, was the life of these vaqueros. Beautiful, too, was their way of expression, their comparison, their philosophy.

Vaqueros were "rough men of color of bronze." To him, they were gallant monarchs, but with the tenderness of children. The vast South Texas brush country—the chaparral—was their palace. They had all the accouterments of a medieval knight, riding a spirited horse on a quest. They rode daily into danger. "Danger, always danger," said Beasley in a recorded interview. In the conflict with their fierce antagonists, the wild steers, vaqueros faced injury and death miles away from home, sequestered in the dense chaparral, alone. As Beasley's mother told him, "the life of the Vaquero hangs by a thread."

Finally, Beasley depicted the vaqueros as modest to the point of humility. Reverent. As they went out into their dangerous work, they did not cheer or howl, but quietly invoked a blessing in their regional Spanish phrase, "En El Nombre Sea De Dios" (In the name of God, with His Blessings). The old vaqueros were saying the same prayer that many Mexican Americans quietly recite today as they embark on an automobile trip on the highway. Indeed, many of the qualities that Beasley bestowed on his vaqueros were hallmarks of the old Tejano community of Texas. The native Spanish and Mexican founders of Texas were "color of bronze," rural, spiritual, and native to the South Texas brush country. They were the descendents of the original Spanish land-grant families like Beasley himself. And for Beasley, the compulsion, the dream was not just the majesty of the vaquero culture, but to reveal this unsung heritage to the modern world. Beasley had his dream. Like his vaqueros, he was on a quest. ★

The Dream

When I was a boy of sixteen, I had a dream by the
fire of a *chamusco*.* We used to build a fire to heat up a blow torch
to singe the spines off the cactus leaf so cattle can eat it during
    droughts.
It was the beginning of the night, and
I had just finished burning prickly pear for the cows.

I built the fire big again to have enough light as
I wanted to sketch the cows around it, eating the prickly pear.
I walked a ways off, sat by a tree and started my drawing.

I had sketched a few cows when, suddenly, I heard
the voice of my heart speak the dream of my life: "Some day
I will make a book to help children." The memory of this scene will
be in itself a reminder of the good faith of that dream, the
guiding light of my life.

This Dream was to bring me the greatest happiness
that no worldly pleasure could have given me; so real it
seemed as if it was a prophetic vision of my destiny.
I felt the confidence in reaching the goal, as if a light,
insight guided me in a world of darkness.

The distance in between me and my goal did not matter. It was not
a barrier; neither were the stumbles in the dark of a lifetime
struggling to succeed.

Chamusco means to singe. South Texas ranchers have
used this modified kerosene blowtorch to singe the
spines on the prickly pear cactus as an alternate feed
source for range cattle when extremely hot weather
dries the natural grasses (drawing 18)

Today, I present you with the realization of that
dream—the book of the life of the old-time Mexican Vaquero
of South Texas—hoping that his philosophy, the drama of his
life, his courage, and his determination never to give up
will create an inspiration for each of you to aspire for a
high goal, to be great but first to be a man, to be noble,
for in the majority of the Vaqueros, to be a man was first in
his code.

I also hope that the great love for freedom of the
Vaquero, of the wild horse, the wild steer, and of the wild
bull in this story will help inspire you with a great spirit
to always stand free.

Whoever you are, wherever you may be, I am with you;
I am one of you. For, by the grace of God, I have never grown
old in my spirit. I am still the ambitious boy of sixteen who
still seeks a goal of helping someone.

My Destiny

In seeking guidance for a spiritual and professional
way of life, we seek light as the plant, born in the dark, does
for existence. So did the Vaquero search for light in his
perils with his only prayer, "En El Nombre Sea De Dios."
In the name of God, with His Blessings.

Heavenly flashes of infinite Wisdom enlighten the
trail even through death. It enlightens the soul, enriches
the spirit to a fountain of joy, transcends happiness, as the
aroma of the amapola beyond its crimson carpet;
and whosoever hears is blessed, and even in death, dies happy.

At the birth of a pale sprig of grass, it is warmed and drawn
through the slightest break in the parched earth
by a thin ray of light that shines upon it,
as if guided by the hand of nature to a better life.
And so it happened to me, that after being given up for dead
when just beginning to live,
I survived through my mother's plea to the Divine Spirit for the
return of my life, as the blade of grass in the shadow of death.

It was soon after this new beginning of life that I
had a ray of light, instinct, to lead my destiny in a way of a
life, as if by mercy.

I will tell you about it, how it began, in a way of song.

As the light of day was dying, a great love was born
that enlightened my life, when for the first time I sighted
a vaquero just out of the brush, coming home by way of the
North Star.

The memory of that vision by twilight is as vivid to me now
As the eternal light of that North Star.
Even today, it seems like the dream of an infant, of a happy awakening.
I can still hear the crickets in the woodpile that night.
And see my mother by my side nearby, dressed in white.
My father just arriving on that little horse he rode—
just a typical mustang, the beige color of the swallows that used to
pause with us on their homeward journey.
This thought frames a picture in my mind of a happy beginning of life.

It was through my father that I had the good fortune
from a very early age to get to know some of the best vaqueros of the time.
When they used to come to visit with him,
they looked so great to me. I admired them so much that
my greatest ambition was to be like one of them, a vaquero.

At my young age at that time, the sense of reason had not awakened yet,
so I had no way of knowing the why of their true greatness,
but instinct must have led my way to have faith,
to believe in and to accept them as my heroes.

Intuitive wisdom and the will of God must walk hand in hand
because they are always right.
The greatness I saw in the vaquero as a child,
and now after my whole life experience with them,
has proven to be true as to the kind of men that they were
and their way of life. I will tell you all about them in my drawings and
     my thoughts.
All that I know.

The vaquero I speak about is the great Texas vaquero of old times,
of this vast region of the brush cow country
between the Nueces river and the Rio Grande
in Southernmost Texas of the North American continent.

I will tell you all the stories of his life of danger, his tragedies, his hardships
when this country was rough, and when the vaquero and
his horse had to depend to the utmost on skill and courage to survive,
as those were the times of the roping of wild steers
and each man was on his own.

It was at this time that a man could be classified as a vaquero
if he was able to rope a wild bull or a wild steer and tie it by himself.
If he couldn't, it was considered a shame for him to claim to be a Vaquero.

The life of the Vaquero began when he was still a kid.
It was just like an old Vaquero, Juan Everett, used to say,
"The making of a Vaquero and that of a cowhorse is the same.
Chasing cows. They share the same passion, the same beginning.
Both have to start young, early like the early spring.
They have to flourish longer, to learn quicker,
and to be absorbed with the spirit of the life.
They must bind with this love forever, for without love,
nothing truly gets complete success."

It was due to their mothers that most Vaqueros had
the good fortune to start young, early in life.
It was because of the mother's loyalty,
to endure the life of peril of their Vaquero husbands,
the hardships, the loneliness of the desolate ranch country.

To the mothers, it was the life of a constant worry,
always expecting the worse, the bad news, the tragedy.
I remember my mother lamented when I was a kid, that
"the life of the Vaquero hangs by a thread."
And she had a right to feel that way.
My father had been nearly dragged to death by a horse
and his body broken up several times during his life.

To the mother was due the place called home, where the
father vaquero came to and where the young aspirant had a
cradle of love to dream that on some happy tomorrow,
he might become a vaquero.
The mother was the life and soul of the home, she gave strength,
her presence fortified the hopes, the ambitions, the dreams.
The father was the leader of that life.
He created the inspiration.

MY GIFT

I have moments of heavenly serenity,
a feeling of an inconparable happiness,
It is at this time that my creative ideas
In my drawings and writing are born.

Thanks God for the pleasure
Of which I am not entitled
To the kind that the world can not give.

I have never been able to
Understand how a sinner like me
Can be the receiver of such a blessing.
Ricardo M. Beasley
February 25, 1962

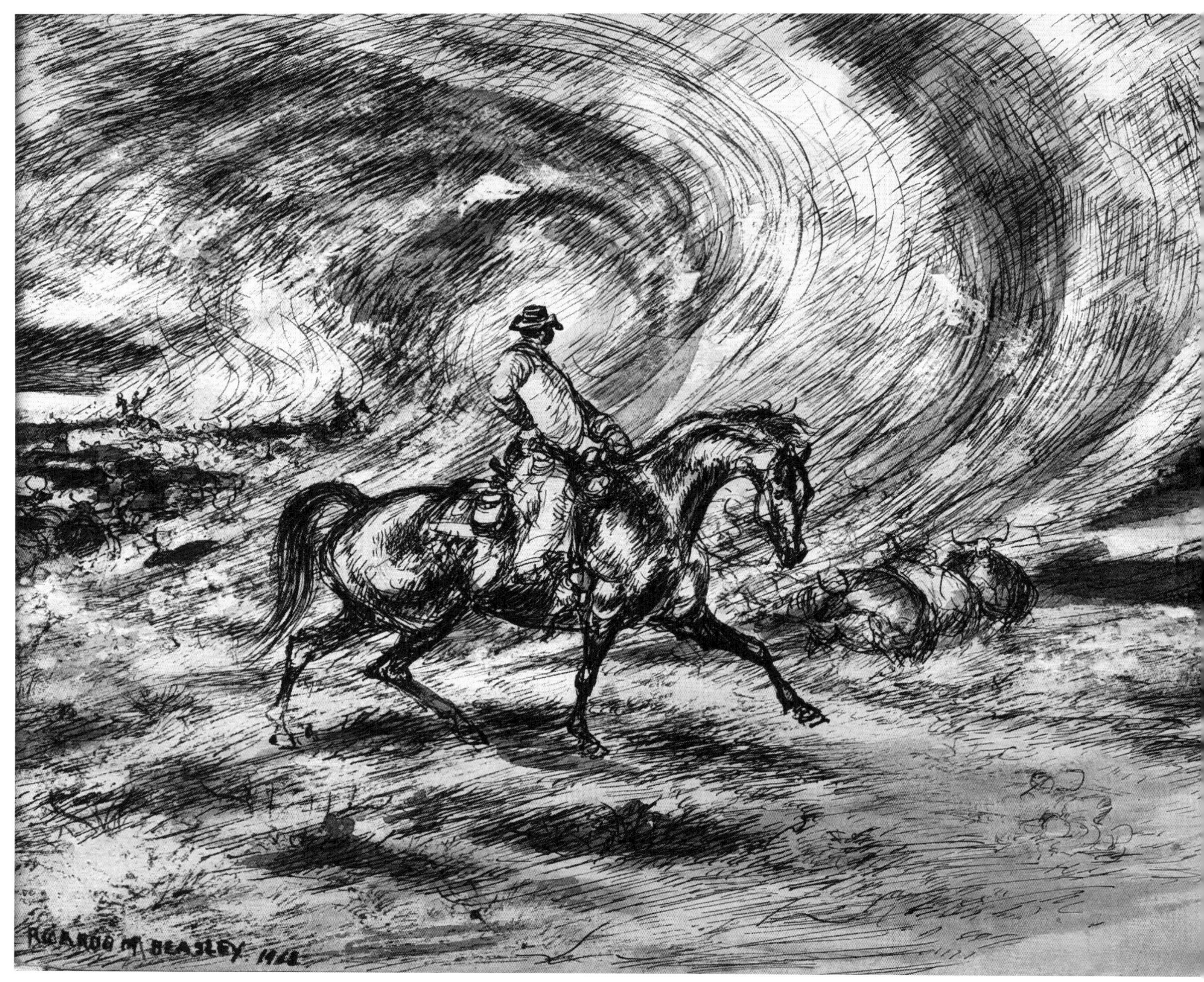

DRAWING 3
Vaquero in a dust storm. © R.M. Beasley
Legacy LLC. Reprinted with permission.

CHAPTER 1

Los Vaqueros

Editor's Note

Ricardo M. Beasley read art magazines, newspapers, and books, but he gained much, if not most, of his knowledge from oral learning and personal experience. He not only listened to the stories around the nightly campfires in the chaparral, but he treasured them. He compared the words of a vaquero's wisdom to the delight of "dividing up a treasure." Throughout his narratives and poems, Beasley attributed his knowledge to the old vaqueros, or he quoted them as references on a wide variety of topics, always with deference to their shared community of wisdom. Throughout his narratives, his pattern was to recount a lesson, and then quote the vaquero's own words. On Juan Everett's knowledge of the wild steers, he quoted the vaquero as saying, "Of course the wild steers were like the Indians—they fought for the right to be free." He quoted another widely known vaquero, Pancho Garza, "El Chinchillo," on the wisdom of silently listening to a person who was speaking, rather than interrupting.

Among the old South Texas vaqueros, there were the legends, and there were the legendary vaqueros. Beasley recounted them both. He cited Marcelo Guzmán, "El Indio," as a vaquero of extraordinary mental powers. Marcelo would lie facedown on the ground with his arms crossed under his face and remain silent in deep concentration. Then he would stand and say, "I already saw the mesteños." He would mount his horse and lead the vaqueros to find a herd of mesteños. The legendary vaquero Sam Smithwick once roped a wild steer by jumping in the fork of a tree and throwing the rope around one branch with a twist of his hand. Marcelo Guzmán once roped two wild steers at the same time, one at each end of the same rope. And once on horseback, he roped a small hawk as it flew off a blackbrush bush. These legends are entirely believable to anyone who has seen South Texas vaqueros, even today, roping a steer by the hind hooves or at the base of the rack of horns.

Another strong characteristic in Beasley's narrative and the conversation of his vaqueros is the anthropomorphosis of the livestock. His narrative frequently refers to the steers as "he," or he attributes human emotions to them. He writes of the love of a vaquero for his horse or the feelings of a horse for his owner. Beasley explains that after weeks in the remote chaparral, a vaquero" starts to see the horses as humans." The horse and vaqero become "friends." When a vaquero is sad, and leans against the corral, his horse leans up to it. "He comes up behind you and nudges you with his nose, he caresses you. He knows when you're sick." And Beasley makes it very clear in one story that a vaquero cannot euthanize his own horse.

A poignant aspect of the vaquero life is Beasley's sad commentary that he woke up one day to realize that vaqueros were all gone. He so feverishly composed his drawing of a *caporal* (foreman) leading his team of vaqueros racing over the hill in pursuit of a herd of wild steers that he did not realize they would never appear there again (drawing 12). They dashed over the hill, and even they were not aware that their days would quickly be gone forever. It was the "unnoticed parting . . . That unrecognized farewell." It was adios to the vaquero. ★

THE VAQUERO had his origin since this was the land of Mexico.
The Vaquero was a typical type of man—his body developed
by the horse and the roughness of that life.
His main characteristics were weather-beaten,
bowlegged, broad shouldered, long backed.
The majority of the vaqueros were of medium stature,
wiry, dark, bronzed-complected. Some were very small men,
but of great endurance like the little horse they rode
of Spanish and Arabian breed.

The life of the Vaquero was hard.
It began at four o'clock in the morning and ended late at night.
Sometimes, especially during roundup times or during rough droughts,
in the scorching heat of summer when they had to burn prickly
pear for the starving cattle, or during the cruel freezes in winter,
a vaquero hardly had any rest.
Their endurance, their loyalty not to give up—not to quit—was admirable.

The dangerous part of that life was more so to the
Vaquero who lived alone in some cow camp where he had to take
care of some big country and do all kinds of dangerous work.
From breaking horses to roping wild steers, he had nobody to
calI on in case of bad trouble—nobody to look for him in case
he was missing, in case of misfortune. He was all alone.
Like my father used to say, "The only ones to find him would
be the buzzards and the coyotes."

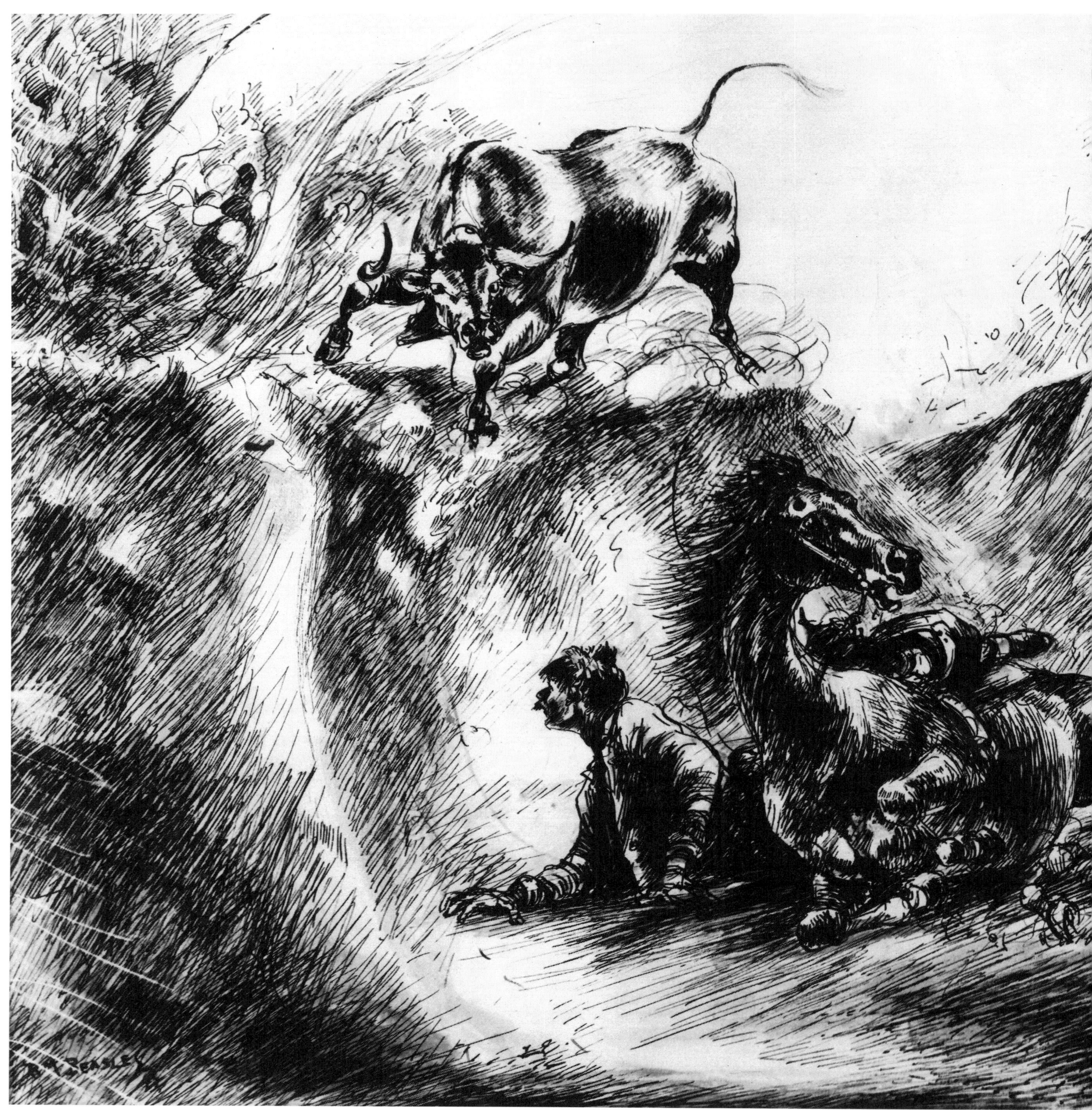

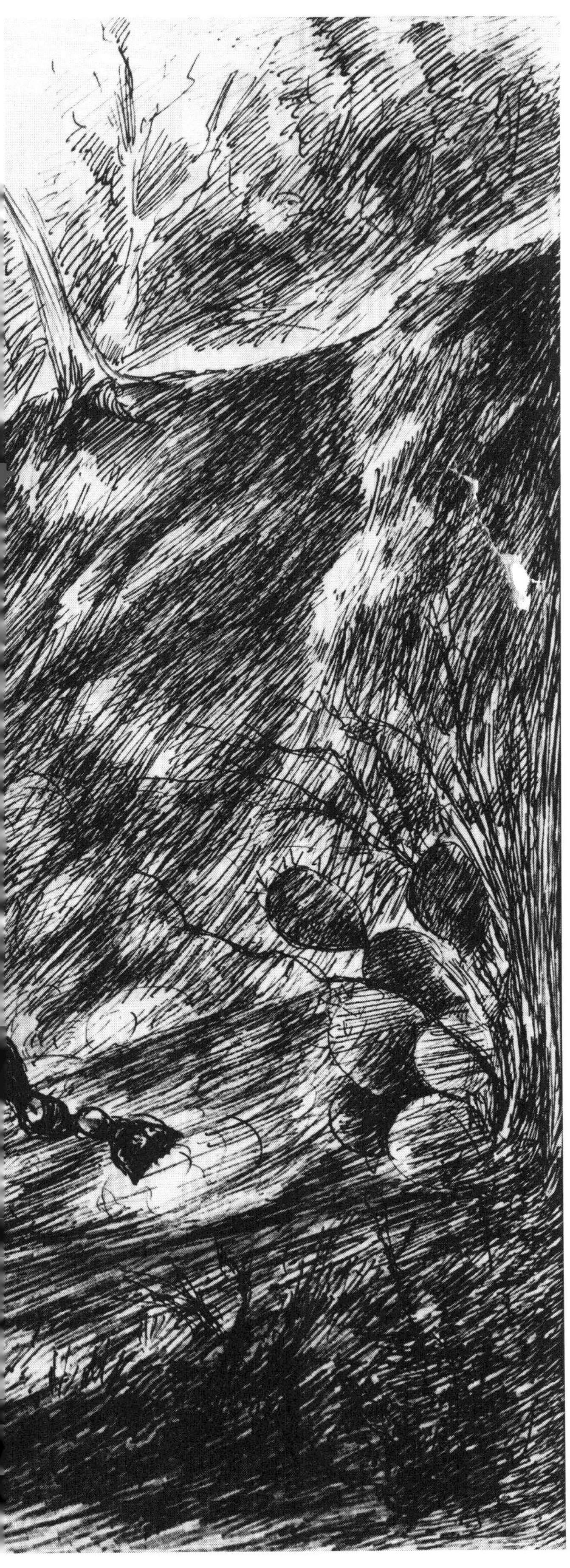

Towards the Infinite Looms the Abiding

Not a cry of help was ever heard from the Vaquero
In the face of the great dangers that existed
When roping wild steers in the night.

Obedience to this decree in the call of duty
Was just another sacrifice in life
He willingly accepted to accomplish his purpose.
To remain silent, with fate alone,
As did Teófilo Salinas that night,
Was the law of the land of the Vaquero.

A cry of help would only banish the hope of success.
Wild steers would thunder away at the warning,
Perhaps, never to return.
    R. M. Beasley

DRAWING 4
"Not a cry for help: Teofilo Salinas that night." © R.M. Beasley
Legacy LLC. Reprinted with permission.

Requiem: Vaqueros

Vaquero—what a beautiful, rich sound that word is to me.
Richer than gold and with a memory as warm as the sound
    of the Christmas Bells of my childhood.
A sound of happiness and joy.

For it seems it was my gift
To learn to love these men that much.
I would give all of the gold of the world
To relive the drama of that life.

I have been fortunate that ever since I was a child
I have lived close to the life that I have loved most
And to the men that played a part to make it shine
—The Vaqueros.

The men of my inspiration for whom I have lived
And whose lives have made mine a happy one.
Today, when the beauty of that life
Has practically vanished, along with the Vaqueros,

We have nothing left but to live in that memory
And enjoy the pleasure in recounting their stories
As if dividing up a treasure.

Rough, too, was the life of these Vaqueros that lived by the chuck wagon
Because they were exposed to all kinds of weather.
These men were weather-beaten too, by the scorching
heat of summer and the freezing winter, but these hardships
did not change their way of life. They loved it as much as a
mother does her child.

I can tell you one story I remember about the the life of a vaquero.
But I don't know if you would want this in English or Spanish.
OK, I could do it the way I want, but I wonder in which would you prefer.
Well, let me just say it whichever way I can.

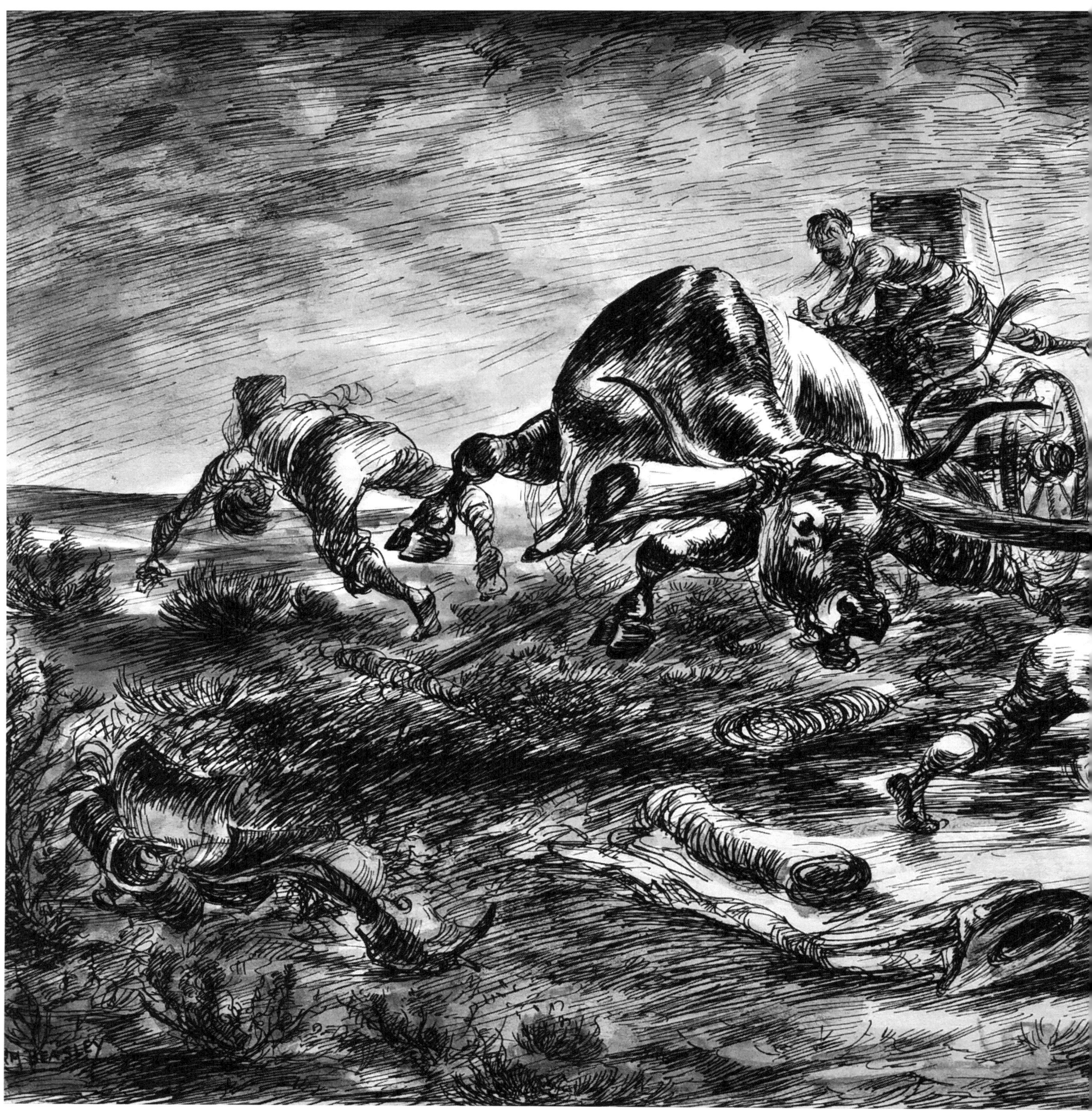

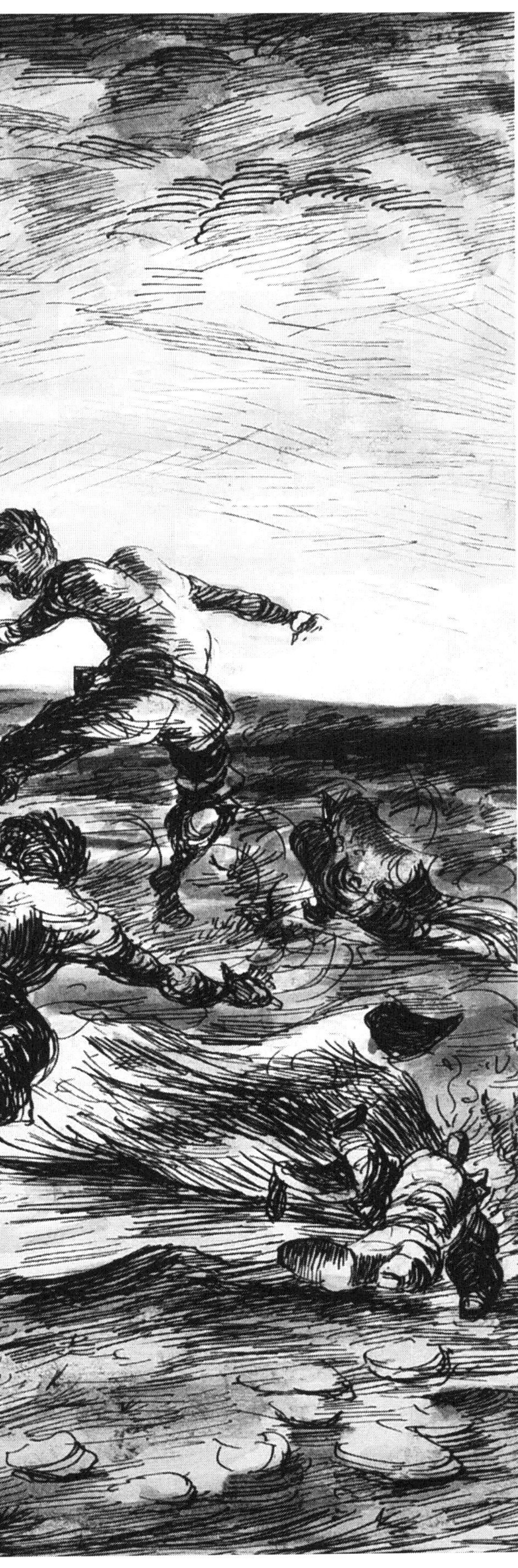

"I REALLY FELT SORRY FOR THE POOR COOK, Pedro Valerio. He was an innocent victim," says Teófilo Salinas. "How come that to happen, I think, was on account of the ignorance of that wild steer. You see, the steer didn't know that the cook is the most harmless creature in a cow camp. The only reason that I can see is that the steer considered him one of the Vaqueros. Anyway, the cook looked like one of them to him, by his point of view, and that was enough to find the cook guilty and sentence him to death."

But Valerio had his justice. Because later that evening, when all the vaqueros had bedded down all around the chuckwagon, another steer attacked the same campfire like a bolt of lightning in the night. The wild steer Geronimo struck the camp of the Vaqueros, to erase from the face of the earth everyone of them for having made a prisoner of him. Their shouts of terror seemed to echo those of the victims of Geronimo, the Apache Chief, on his raids of devastation. And surely the intentions of Geronimo the steer were just as deadly as those of Géronimo the Chief.

This is the same bunch of Vaqueros that laughed so much at the poor cook cowed at the top of the chuckwagon box by the wild steer early that day. The only difference now is that they have lost all the feeling for what they previously thought was entertainment, but now they cannot afford to laugh, not even smile. At the speed they are scattering, they couldn't even think of it.

The only one to enjoy a good hearty laugh and really enjoy the show this time was the cook. He could afford it. You see, by the laws of the cow camp, the cook is the only one that has the right to sleep in the wagon. The other vaqueros were bedded down all around the camp.

DRAWING 5
Like a bolt of lightning the wild steer Geronimo
struck the camp. © R.M. Beasley Legacy LLC.
Reprinted with permission.

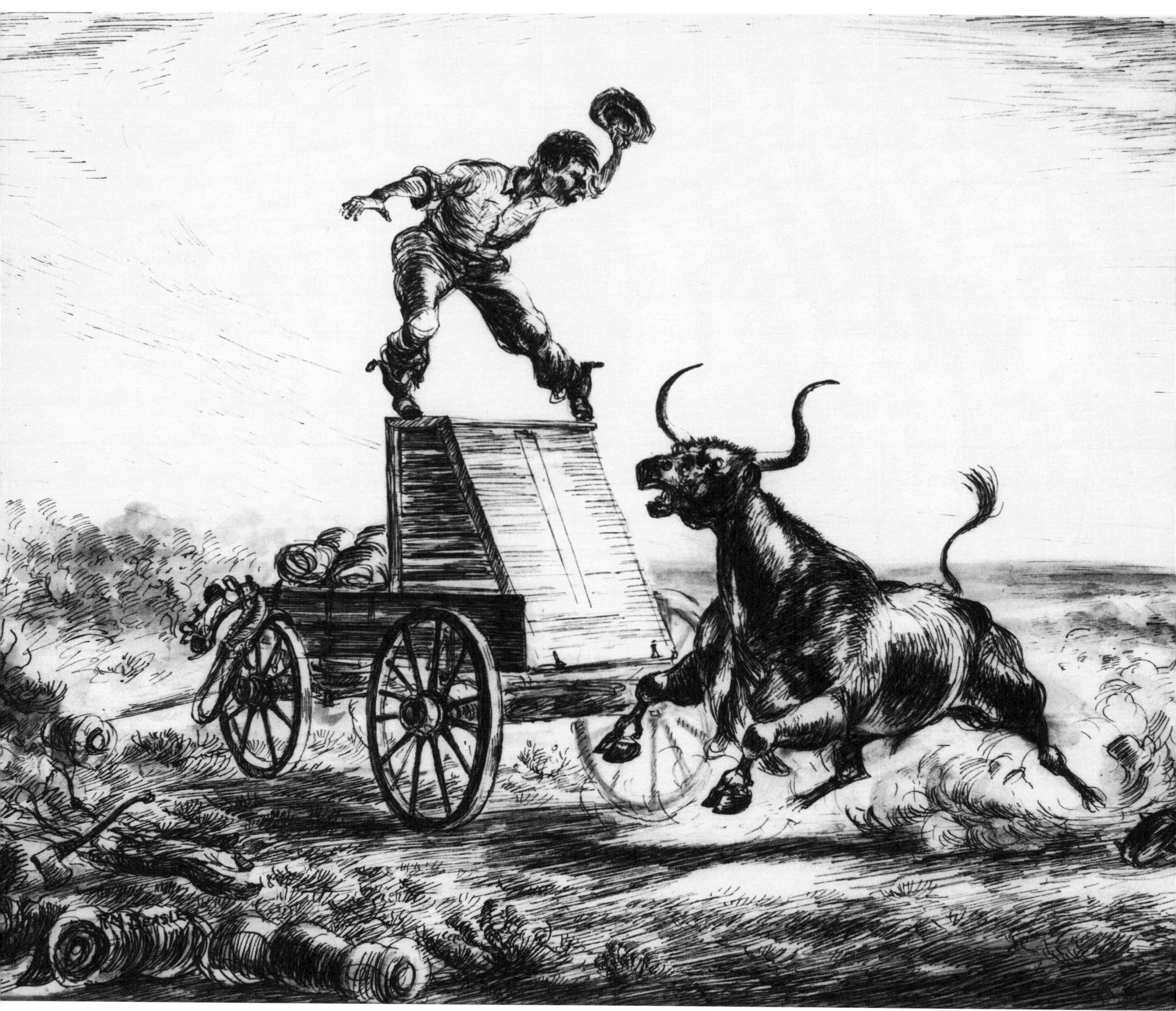

The Innocent Victim

CUANDO ANDABAN estos vaqueros en su trabajo, que en veces andaban fuera de sus casas, pero en las películas de hoy en día y todo eso, y casi una vida lujosa pero no era así. En verdad, estos vaqueros traían lo menos posible porque andaban en caballo. Sí. Es verdad.

Cada quien tenía su—una colchita, algo para . . . bed roll. Una lona. Usaban un tarpolio. Una lona gruesa, era, y allí traían su camita. La lona era como está hay en el dibujo de los vaqueros dormidos, esperando los novillos ladinos. Como dormían afuera, llovisnaba y se la hechaban arriba.

¿Y de comer? Uno cree que tampoco, muy poco que llevaban con ellos. Pero no es verdad.

Todo el tiempo traían chuckwagon. Cuando andaban trabajando duro, eso, cuando andaban de caminantes nomás de un lugar a otro, traían nomás lonche poquito para ellos, pero cuando andaban trabajando, traían su chuckwagon box. Bastante comida. Bastante.

Nomás que tenían que dormir afuera.

Ja ja!

I remember one time—a tale—I want to relate it, but I don't want it to come out in a book. I will tell you another one about a cook.

Esta historia que es aquí en el condado de Duval al lado de Siete Hermanas. Hay un rancho que se llama Eagle Hill Country. Foster. Foster Ranch.

Y tenían un ganado tan salvaje en ese rancho—quien sabe que clase de cría sería, pero eran. . . . Anyway, esos novillos ladinos que tenía, no los podían pescar.

Se hicieron salvajes porque en ese tiempo los Foster no vivían en rancho. Estaban solos. Se hacían hay mesteños, pero lo raro era lo peludo que parecían leones (drawing 7). Tenían clin y una llave como de toro de pelea.

ONE THING THAT I've always wanted people to know is that when those vaqueros were out doing their work, out at times away from their homes, but in the movies and all that it's almost a luxurious life, but it wasn't like that, really. These vaqueros took the bare minimum because they were riding horses. Yes, that's right.

You each had your—a thin blanket, something to . . . a bed roll. A canvas. They used a tarp. Just a thick canvas it was, and it held their bed. The canvas was as it shows in my drawing of sleeping vaqueros waiting for the wild steers. Because they could sleep outdoors, in the mist, they held it over their head.

And to eat? One may think that they didn't take much with them, but that's not right.

No. They always took a chuckwagon when they went out on a big job, like that. Now when they went on the road from one place to the next, then they took only a light lunch for each man. But when they were out on a big job, they took their chuckwagon box. A lot of food. Plenty.

But they still had to sleep outdoors.

Ha ha!

I remember one time—a tale—I want to relate it, but I don't want it to come out in a book. I will tell you another one about a cook.

This one story that happened here in Duval County next to the Seven Sisters at a ranch that is named Eagle Hill Country. Foster. Foster Ranch.

And they had such a wild livestock on that ranch—I don't know what breed of stock they were, but they were . . . Anyway, those wild steers that they had, nobody could catch them.

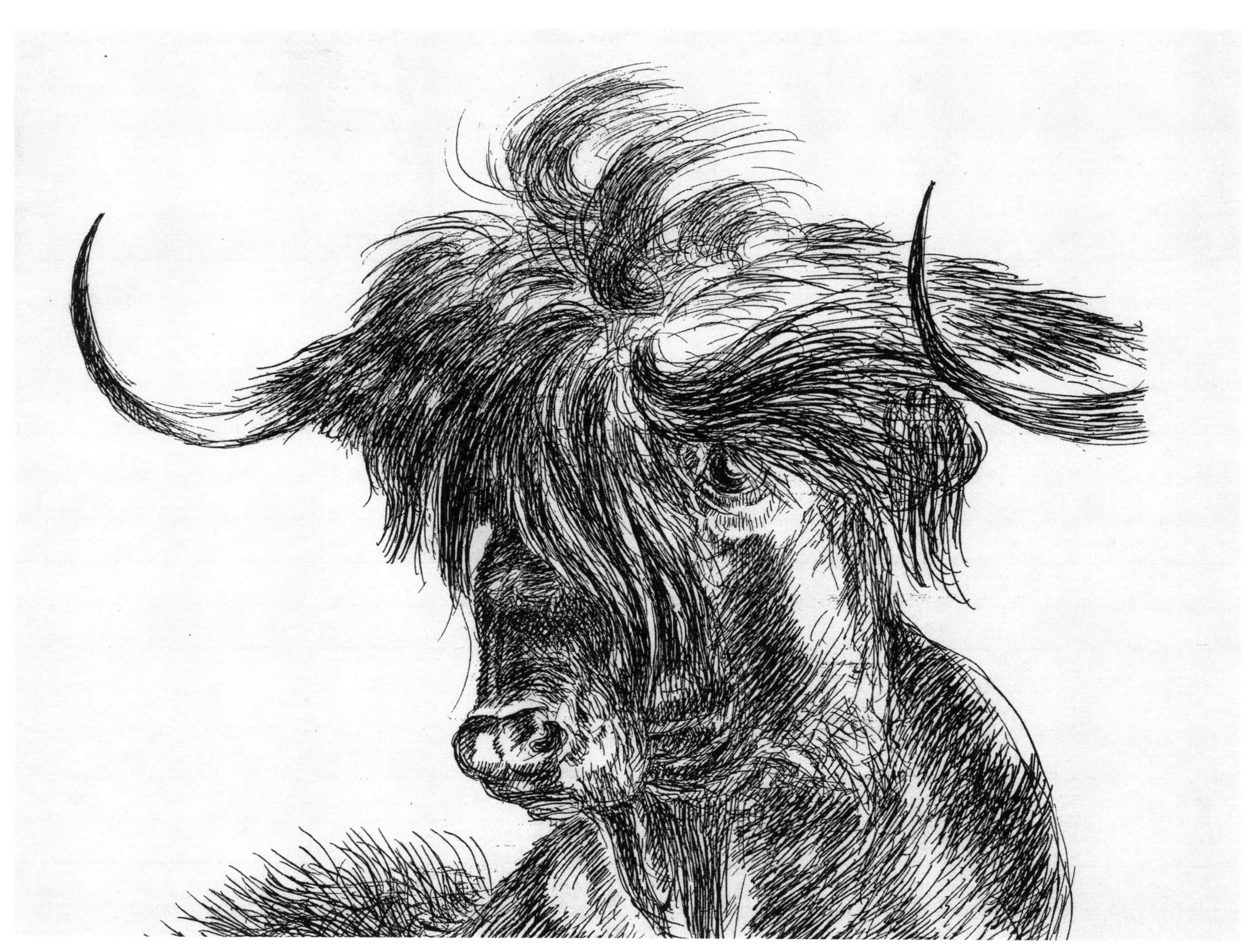

Esos animales eran tan brutos, tan ladinos que cuando los iban a arrear que los tenían ya pescados en una trampa, de aquí pa' allá iban detrás de los vaqueros. Llegaban a la otra cerca, y luego los novillos traían a los vaqueros pa' atrás. Ja ja ja ja.

Había un comprador aqui de ganado, Skidmore, que viene siendo el papá de Harry Skidmore, y él le compraba el ganado ladino a los Foster—lo que pudieran pescar—lo que quedaba ladino, y el llevaba sus vaqueros y todo.

Me platica a mí una de las muchachas de Foster porque todas son muy vaqueras, muy trabajadoras. Dice que esa vez traían un ganado, de allá de Siete Hermanas. Ellos tienen un rancho aca en Palito Blanco, y traían un atajo de novillos, un ganado pa' aca. Y durmieron en La Gloria. Hicieron campo en La Gloria. Y dice, "En la mañana, mis hermanos," dice, "ensillaron y fueron a dar la vuelta al ganado a ver si no se había ido alguno, un animal, y pescaron la huella de un novillo que se había ido. Y lo siguieron, fueron y lo lazaron, y lo trajeron pa' tras, y eso es cuando lo soltaron allí donde estaba el cocinero."

Y ese día tenían su campito allí en el potrero, un guayín, y lo tenían en una trampa en un potrero chiquito, y estaban lazando los novillos en un potrero grande, y los metían allí.

Y esa tarde se trajeron ese novillo prieto, y lo soltaron en la trampa pa' él, y naturalmente agarró la cerca el novillo buscandose salida, en el rumbo que lleba. Estaba en una trampa chiquita, que había una trampa el campo de Skidmore.

Y el cocinero está dando providencias hacer de cenar cuando el novillo enfrenta y lo ve. Y ya venía enojado, y que entre el cocinero. El cocinero arrancó, y estaba la cafetera arriba del respaldo de las brazas, llena de café. El cocinero ganó por el guayín.

Y vino el novillo y dio una vuelta a la cafetera con la nariz, y lo tumbó. Y onde calló el café en las brazas, el vapor le quemó la cara y peor, y que coraje le acusó al pobre el cocinero que fué y lo sacó de bajo del guayín. Se lo quería voltiar al guayín (drawing 6).

Se subió arriba y lo alcanzaba, al cocinero, al cajón. Ya no tenía. Ya no tenía pa' onde hacerse, y empezó gritar "ayuda."

They became wild because in those times the Fosters didn't live on the ranch. They were left alone. They became wild, but what's rare is that they were so hairy that they looked like lions (drawing 7). They had a rack of horns like just a bullfighting bull.

Those beasts were such brutes, so wild that when they were driven and already caught in a pen, going from here to there, they would go behind the vaqueros. They'd get to the other fence, and then the steers would bring the vaqueros right back. Ha ha ha ha.

There was a livestock buyer here, Skidmore, who is the father of Harry Skidmore, and he used to buy wild livestock from Foster—as much as they could round up—all the wild livestock that was left—and he would take his vaqueros and all their equipment.

One of Foster's daughters told me that once that— because the daughters were all good vaqueras, hard working. She said that day they were bringing that herd from over at Seven Sisters. They had a ranch over here at Palito Blanco, and they were driving a string of steers over here. A herd. And they slept at La Gloria. They camped out at La Gloria. And she says, "The next morning, my brothers," she says, "saddled up and went to make the rounds of the herd to see if any of the, any beast had escaped, and they caught the trail of a steer that had gotten out. And they followed it, and they went and roped it, and brought it back, and that's when they turned it loose over where the cook was."

That day they had their trap set in the horse pasture, a wagon, and they had a pen there in a little horse pasture. They were roping the wild steers in the large pasture, and they would put them in it.

Well, that afternoon they brought in that black steer, and they put it in a pen for itself. Well, naturally the loose steer set about looking all around for an exit, in the direction he wanted to go. It was just a little temporary pen that was in Skidmore's pasture.

Well, around dinner time when the steer approached and saw the cook and the camp fire. Well, it was already furious, just as the cook entered, so it goes straight for the cook. The cook fled. The coffee pot was set on a rack on top of the embers, full of coffee.

Y los vaqueros estaban allá en el corral donde habían echado el novillo. Se vinieron a los gritos de él.

Y está es la ultima punta, ya no podía subir más. Ya no tenía pa onde.

Everybody thought it was funny but the cook. He had a right not to think so. It's not funny at all to be in a position close enough to hell to smell the smoke and not an inch of room to back away.

Ja ja.

Me acuerdo del nombre del cocinero ese.

Pedro Valerio.

Valerio.

Pedro Valerio. Sí.

El novillo era prieto. Novillo prieto. Les decían los vaqueros "horribles." Parecían . . . parecían leones (drawing 7). Tenían pelo largo en la cabeza, en la frente, que tenían que sacudir la cabeza pa ver. Brutos salvajes. Bueno que tal peleaban con la lumbre.

Peleaban con la lumbre. They remember it. Cuando los marcaban de chiquitos, este, se levantaban y le caían arriba de la lumbre y las brasas que saltaban las seguían como canicas desparramadas todas. Y otra cosa, ya de grande, nunca se les olvidaba esa lumbre a ellos. Donde había lumbre, iban y lo apagaban.

Just as the cook dived under the wagon, the steer came and turned the coffee pot over with its snout, knocking it over, and as the coffee spilled into the embers, the steam burned its face. Even worse. It directed such fury at the unfortunate cook that it went and got him out from under the wagon. It started trying to upend the wagon (drawing 6).

The cook climbed on top, and the steer stretched upward, tring to reach the cook on top of the chuckwagon box. He had no place to hide. He had no place else to move, so he began screaming "Help."

The vaqueros were in the corral where they had first brought the steer. They came to his screams.

He was at the highest point and could climb no higher. He had no place else.

Everybody thought it was funny but the cook. He had a right not to think so. It's not funny at all to be in a position close enough to hell to smell the smoke and not an inch of room to back away.

Ha ha.

I remember the name of that cook.

Pedro Valerio.

Valerio.

Pedro Valerio. Yes.

The steer was black. A black steer. The vaqueros would say they were "horrible." They looked like, like lions (drawing 7). They had long hair on their head, on their forehead, so long that they had to shake their head to see.

Savage brutes. And how they fought with fire.

They hated the campfire. They remember it. When they were branded as little calves, well, they'd jump up and stumble over the fire, and the embers that flew up, they'd chase them like marbles scattered all about. And another thing, once they grew up, they never forgot that fire. Anywhere they saw a campfire, they would go and stomp it out.

Vaquero Wisdom

The majority of the Vaqueros were good men. They
seemed to live by a code of tradition, to be quiet, formal,
modest, and honest. The honor of duty came high above all;
they were possessed of a great determination. An old man
that was raised with the Vaqueros once told me, "To be a
Vaquero, a man had to be brave and intelligent."

The Vaquero survived hardships, injuries and illnesses largely
because of his knowledge of nature's medicinal herbs. He knew
the plants, barks, roots, the hearts of certain trees that were
good for one disease or another. They were self-sufficient all
the way around. Some of their medicines were used to treat
injuries on their horses also, and the effect was just as good
on themselves.

To the Vaquero, the worth of a man was not based
on how much money or knowledge he had, but on his kind of word,
the good word, was their seal of honor.
My father used to preach to me since I was a child,
"Never speak more than what you can stand for."
Advice that I still follow.

The Vaquero had a deep faith in the supreme being.
All their work was done in the name of God,
and if they secceeded in some work
or got out of a great danger alive while roping some wild cattle,
they never attributed it to their skill or courage,
but they would say, either, "God helped me,"
or "If I am alive, it is because of an act of God."

The Vaqueros were not men of education,
 but were men of respect, kind and very courteous.
They had a culture of their own.
They had no connection with either the Mexican or Anglo culture.
Their own standard was based on sound traditions, principles that
     make a man.

They were eloquent in their speech, like the Indian Chief,
very reserved, great observants, good listeners, great
     conversationalitsts.
My father used to advise us, "If you don't have something
     interesting to say,
don't interrupt, listen, so you may learn."

The best example of listening I learned from an old
Vaquero, Pancho Garza, "El Chinchillo." One day he and I were
visiting with an old Vaquero friend of ours, Alberto Vela.
And all the time we were there, Alberto did the talking,
reminiscing about old times of the Vaquero. Pancho never
said a word, just listened intently (drawing 8).

After we left there, on the way back to the ranch, I asked Pancho,
"As much as you know about ranch life, why didn't you tell some of
     your stories like Alberto did?"
And he said, "Why interrupt the man?
He felt like talking, and he knew what he was talking about.
 It was interesting." It was a custom among the Vaquero
to be silent for hours if he didn't have something of interest to say.

Among the Vaqueros were men of wisdom, of rare
abilities, deep insight, very observant of nature. Many of
them might not know how to read, but they could read your mind
and tracks on the ground like an open book. Pancho was one
of these rare men; he was possessed with the rare gift of
the sixth sense, he could foretell a lot about dangers ahead.

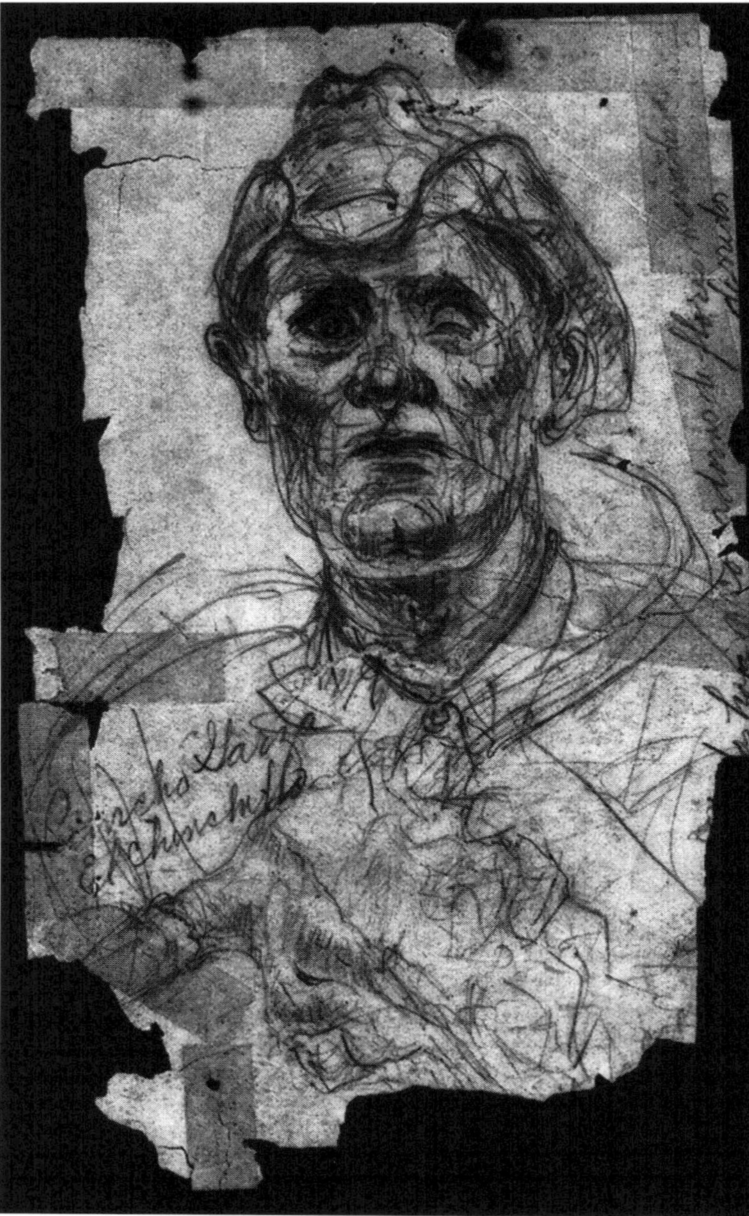

DRAWING 8
"El Chinchillo," Pancho Garza.
Drawn impromptu on grocery
paper bag. *Courtesy of Manuel Peña.*
© R.M. Beasley Legacy LLC.
Reprinted with permission.

Speaking of observances of nature,
I learned something unusual from a King Ranch Vaquero,
a Kineño named José Lopez.
One day he and I were riding through some open country
where the brush had been knocked down. Suddenly he said,
"The country used to be more healthy when there were woods;
now all we breathe is dust. I figured out what he meant to say,
was that the oxygen is what we missed from the green vegetation.

All these Vaqueros were men of their own,
of a deep self-confidence. Pancho would say that education didn't
mean much if a man couldn't have something of his own.
And to excel in something distinguished was not a simple act.
But achieved with the confidence you have in the ability that God has
given you. So it was the same in reaching the goal of any
pursuit. Self-confidence is the bridge in between, and he
would add, "Have confidence and never fear and you will always
succeed, because if you are afraid, you will never get anywhere."

From Pancho I heard the best definition of the vainess that is a lie.
"A lie," he would say, "is like the fire, or the water, or the wind;
it has no hold." He would also say that whoever is good to his mother,
God would always help him.

Marcelo Guzman was another Vaquero of rare mental
powers. He said that during the time of the mesteños—the mustangs
that ran free across the land—he was searching for them once, and
unable to find the herd. Marcelo got off his horse and told his companions,
"I will tell you in a little while where they are."
Laying face down on the ground with his arms crossed under this face,
he would remain silent in deep concentration for a short while;
then, he would get up and tell the others, "I already saw the mesteños."
And getting on his horse, he would take the lead and find the herd.
On account of that ability, he got to be called, "El Indio" Marcelo.

Beautiful, too, was their way of expression, their
comparison, their philosophy. An example of this was Juan Everett.
One day I was telling him that in my drawings,

I always give the wild animals the upper hand because they are the only
ones in the right. "Of course," said Juan, "the wild steers
were like the Indians—they fought for the right to be free
and to hold their country."

To be a true Vaquero was considered to be fulfilling
the highest endeavor of man; however, the best of the Vaqueros
never claimed to be a Vaquero, for it seemed to them that
to be a Vaquero was a goal as much of an impossibility
like reaching the top of an inaccesible mountain.
Because, as they would say, "The work of the Vaquero is never learned;
the best fail, the best get killed."

The Vaquero would avoid the use of the word "I"
as much as possible; as if to speak of oneself was a major sin.
He prefered to speak in the second or third person.
When and if by some necessity he would speak of himself,
he first would apologize by saying, "It is bad that I say this,
that I should speak of myself, but I will say it because
the occasion comes up or because you asked me."
And then, only if the case was such that he had to speak of himself.

In the cattle industry in this part of the country,
the Mexican Vaquero of old times played the major role
in the handling of the cattle.

The modern cowboy is the offspring of the Vaquero,
and to the Vaquero is due the taming of the once-wild cow country.

This was the kingdom of the Vaquero; he was supreme.
There will never be the likes of them again.
The times that made them great, that kind of man,
that unique breed of Vaquero is over.
There is hardly anything wild left to speak of anymore in this country.

I miss the dying howls of the Vaqueros from afar
in the chase of some wild steer, as if going into eternity.

The memory of it, now, seems to me, the vanishing echo of
a life in agony. When the roar of the lion is no longer
heard, the King of the forest no longer exists.

The Vaquero has passed away, but not in vain. He lives in our hearts
as an emblem of honor and pride, that once existed, but is no more.

For the Love of His Horse

The horse and the Vaquero, they become each other's friends.
I knew a story about a horse that the vaquero rode it without even
knowing which way the horse was running until he

felt the wind hitting him on the face. I did a drawing that
showed how much the Vaquero loved that horse, and
I can tell you about this one.

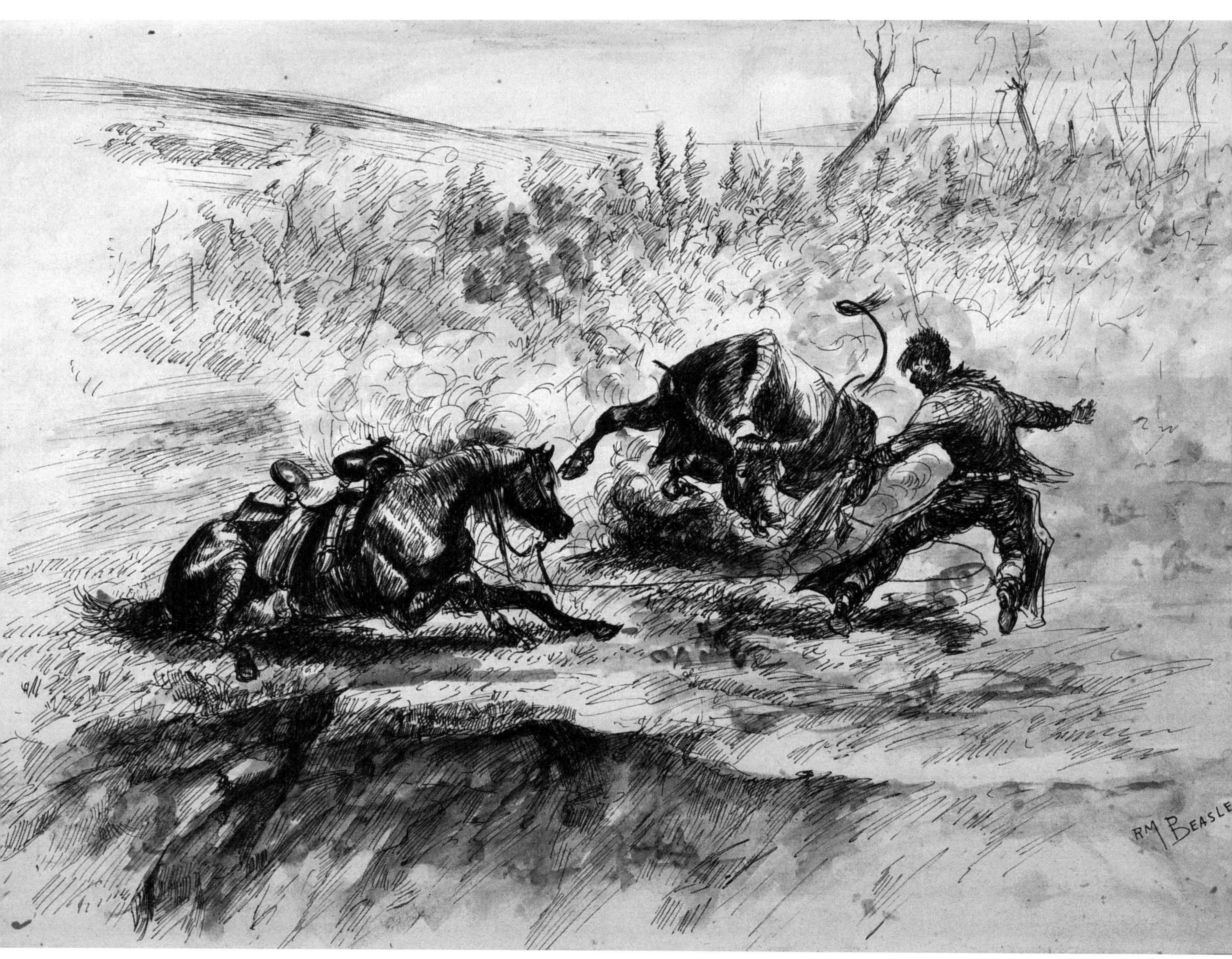

DRAWING 9
"Lo toreó . . . Para cansarlo." (He bullfought it
. . . to tire it out.) © R.M. Beasley Legacy LLC.
Reprinted with permission.

ESTE, ME PLATICABA el señor, era americano el que tenía un café aquí en San Diego en ese tiempo. I forgot his name. Dossgard. Dossgard. Dice que andaba trabajando. Lazó un novillo que era de Dobie, el novillo ese. Y al tiempo que lo lazó, brincó un arroyo, y brincó el caballo, y lo pescó en el viento y lo tumbó. Y le quebró una pierna. Danza Doche se llamaba el caballo.

Sí, Danza Doche. Y entonces, él, para defender el caballo que no lo matara el novillo, se quitó la chaqueta y lo anduvo toreando hasta que lo cansó. Y luego, bueno, he bulldogged it. Increible, pero primero lo toreó . . .

Para cansarlo. Para cansarlo (drawing 9).

Lo dibujé, pero no me dí cuenta del año. Haya haber sido in the twenties, por allá porque ya para los treinta, no había mucho ese costumbre hay de ganado, pero ha de haber sido esos años. Él conoció a Sam Smithwick, a todos los vaqueros buenos, Ambrosio Montes.

It was one of my first action drawings.

And I marked it with the date, March of '46.

Y el caballo, lo quería tanto que mandó a otro que lo viniera a matar. They had to kill it. No tenía remedio. Alberto Lazo fué el que vino a matarlo.

El mismo vaquero no quizo matar su caballo. Well, you can understand that. Yeah. Loved his horse.

Cuando vive uno en un rancho con un caballo que está uno solo, a los caballos, los vas viendo como humanos. Se hacen, se hace uno amigo uno de otro. Y el caballo es muy inteligente. Yo sé, estas triste, el corral, recargado. Viene por detrás y te hecha el pico, te hace caricias. He knows when you're sick.

WELL, A MAN TOLD ME, he was an Anglo American who used to own a café here in San Diego. I forgot his name. Dossgard. Dossgard. He says he had lassoed a steer that way. It was one of Dobie's. So, as he lassoed it, it jumped over an arroyo. And just as the horse jumped, the steer caught the horse in midair and threw it to the ground, breaking its leg. The horse was named Danza Doche.

Yes, Danza Doche. So in trying to protect the horse from being killed by the steer, he took off his jacket and did bullfighting with it until it got tired. And then, he bulldogged it. Hard to believe, but first he was bullfighting it . . .

To tire it out. To tire it out (drawing 9).

I did the drawing, but I didn't note the year. It must have been in the twenties, around then because already by the thirties, there wasn't much of that herding left there, but it must have been in those years. He knew Sam Smithwick, all the good vaqueros, Ambrosio Montes.

It was one of my first action drawings.

I later finished it, and marked it with the date, March of '46.

And that horse, he loved that horse so much that he had to have somebody else come kill it. They had to kill it. It was beyond healing. Alberto Lazo was the one who came to kill it.

A vaquero wouldn't want to kill his own horse. Well, you can understand that. Loved his horse.

When somebody lives on a ranch with his horses and he lives alone, he starts to see the horses as humans. They become, they become each other's friends. Because the horse is very intelligent. I know, when you're sad, the corral, he leans up to it. He comes up behind you and nudges you with his nose, he caresses you. He knows when you're sick.

The Vaquero Spirit

The Vaquero was known for his ability to race through
the brush behind some wild cattle at the full speed on a horse.
Some were so good at it that they would do it without wearing
a dust jacket and their shirt would not be shredded off by thorns.
Others were so daring that they would take the bridle off their
horses while on the chase of some wild steer and nothing
would stop them but the Steer at the end of the rope.
Of this kind I knew three Vaqueros, Colon De La Garza,
Sam Smithwick, Jr. and Pablo Rangel.

Marcelo Guzman and Sam Smithwick were known
as artists with the rope. Sam once roped a wild steer by
jumping in the fork of a tree, and throwing the rope
 by one side of the fork with a twist of his hand to land the rope
firmly on the steer.

Marcelo Guzman was so fast with a rope that it was
amazing. During the time of the mesteños, he was known in
his country as the best roper among all the mustangers.
He once roped two wild steers, one at each end of the rope,
practically at the same time, and let them go to tangle up in the thicket.

Once I saw him rope the hind hooves of a Brahma steer
in the flash of a second, as the steer spun around at full
speed at the gate of a corral. What amazed me besides his skill
to throw the rope, was his fast thinking. To think about it,
in such unsuspected instant. There was a time, on horseback,
when he roped a small hawk as it flew off a blackbrush bush.
The memory of that still amazes me.

The ability of the Vaquero to tie down wild cattle
after roping them in the brush, was unique. They had a wisdom;
even so far as to tie down a wild steer with its own tail or
to tie a steer to a tree with the saddle horn string.
Interesting, too, was their variety of rope ties that they
used to tug the wild steers out of the thick brush.

The greatness of the old Vaqueros was not their skill.
It was their spirit, their innocent love of the chase, and their loyalty.
I will never forget seeing their gratitude to God and to each other
There's a long story I can tell about Rodolfo Muñoz and Sam Smithwick.
It's a long, winding story—with a sad ending. *Triste y penoso*. Sad and painful.

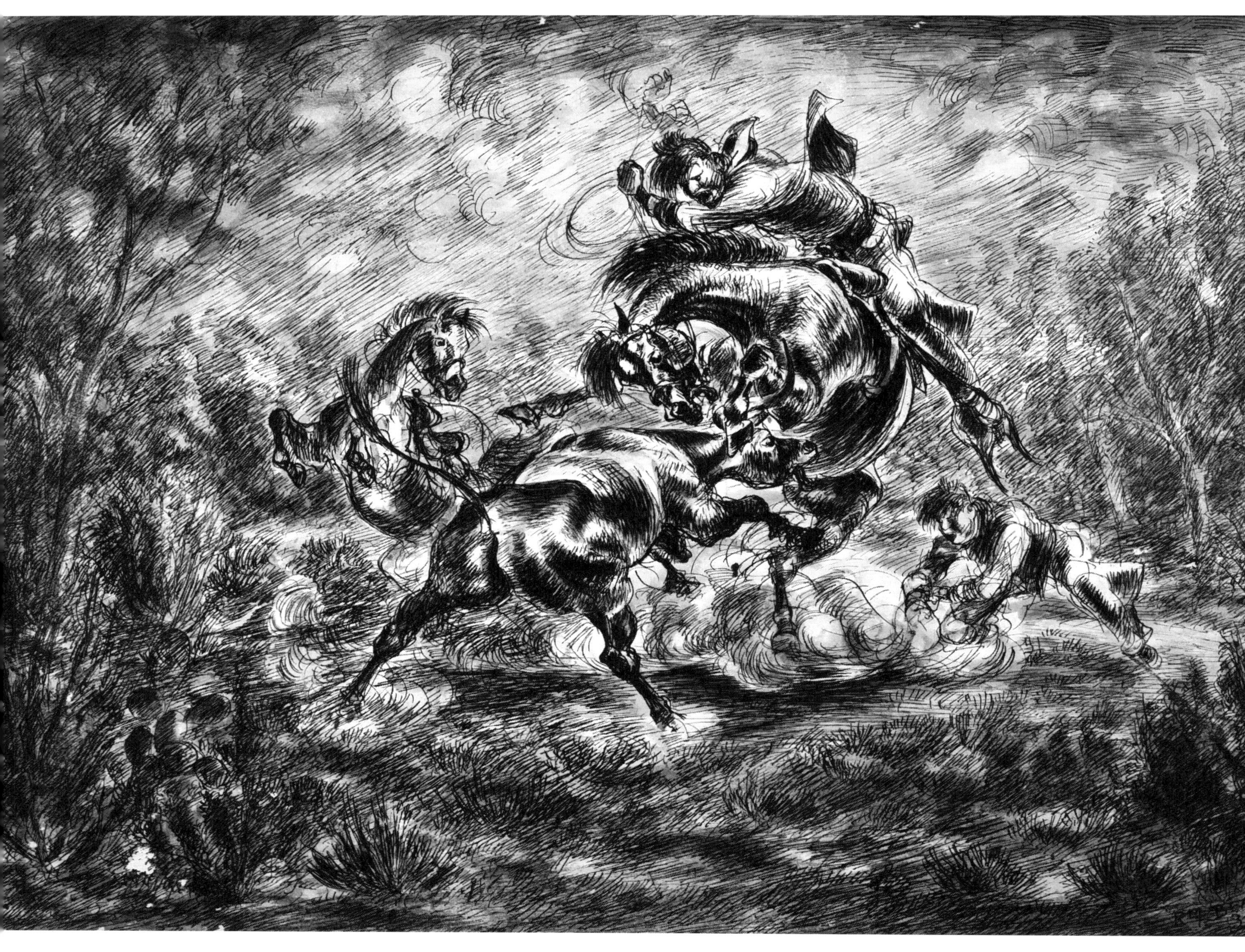

DRAWING 10
"Santiago saved my life by dashing his horse at the extreme
speed into the horns of the steer." © R.M. Beasley Legacy LLC.
Reprinted with permission.

The Reverence of a Vaquero

ESE ES MUÑOZ en este dibujo. Es Rodolfo Muñoz, el hijo de Santiago. Santiago es el del otro dibujo en que le salva la vida a Sam Smithwick (drawing 10).

Yo creo que ese día que estuvo aquí en San Diego la esposa de Tommy Molina. Yo le platiqué yo que iba yo allá en la botíca de Donato cuando Muñoz venía de la otra calle y me hablo que me parara pa' contarme la historia allí mero, en frente de la botíca. Y loco de gusto porque se había escapado que el toro no lo matara porque los perros se los quitaron de en cima (drawing 11).

Pero el toro había atacado. Solo quiso lazarlo.

Ese fué en La Juana. La Juana Ranch.

Sí. No, lo andaban juntando. Lo lazó al brincar el nopal el caballo. Y le tumbó el caballo, y se lo llevó en rastra. Y llegaron los perros y lo distraeron más, y se le reventó el mecate.

Estos vaqueros, a veces usaban bastante esos perros.

Algunos. Los viejos no usaban casi nunca el perro.

Pero si vieras de la manera rara que supe yo esa historia. En novecientos cuarenta siete, no sé si acordará la gente. No se pueden acordar de eso. Hubo un tiroteo allí en la botíca de Donato y mataron al hijo, el papá de Raúl, y hirieron al viejito. Los Everetts, los Everetts.

Y cuando estaba en el hospital el viejito, va a verlo Sam con Willie Mancha, otro amigo. Y cuando lo vió Sam en la cama al viejito herido, se puso Sam a llorar.

Se soltaron las lagrimas.

Y le dice al compañero, "Yo le devo la vida a este hombre."

THIS IS MUÑOZ in this drawing. It's Rodolfo Muñoz, the son of Santiago. Santiago is the one in another of my drawings where he saves Sam Smithwick's life (drawing 10).

I think that was one day Tommy Molina's wife was here in San Diego. I related to her that I was going over to Donato's drug store when Muñoz was coming from the other street, and he called to me to stop so he could tell me the story right there in front of the drug store. Overjoyed that he had escaped being killed by the bull because the dogs had drawn it off him (drawing 11).

The bull didn't really attack. He tried to rope that bull. That incident was in La Juana. La Juana Ranch.

Actually, they were going to bring in the bull. He lassoed it just as the horse was jumping over the cactus plant. It pulled the horse down and dragged it off. That's when the dogs came up and further distracted it, and the rope broke.

Some vaqueros used those dogs often.

Some. The old vaqueros hardly ever used a dog.

But it's just the unusual way I got to know that story. Back in 1947, I don't know if many people remember. They can't remember that. There was a shooting over at Donato's drug store and they shot the son, the father of Raúl, and they wounded the old man. The Everetts, the Everetts.

When the old man was in the hospital, Sam goes to see him with Willie Mancha, another friend. But when Sam saw the old man in the hospital bed, wounded, Sam started crying.

The tears flowed.

And he tells the other friend, "I owe my life to this man."

Y ya le cuenta la historia con eso, como pasó, y por qué razón. Y cuando le platicó a Willie lo que, la historia, entonces el viejito le voltea a Sam. Y le dice a Willie, "Es cierto así como te contaron."

Y tenía, la palabra que puso él, dice al ultimo, "Muy valiente Santiago," dice, "Muy pronto para pensar."

Y le dice como todo eso fué en un segundo. A flash. Así como caió, pasó sobre él, y topó el novillo.

With that, he tells him the story, how it happened, and why. After telling Willie what happened, the whole story, then the old man looked up to Sam, and told Willie, "It's true just as he told you."

Sam had a way of saying it, the way he worded it, he finally says, "Very valiant, Santiago," he says, "Very quick to think."

And he told how all that happened in one second. A flash. The way he fell down, how it flipped over him, and the steer butting him.

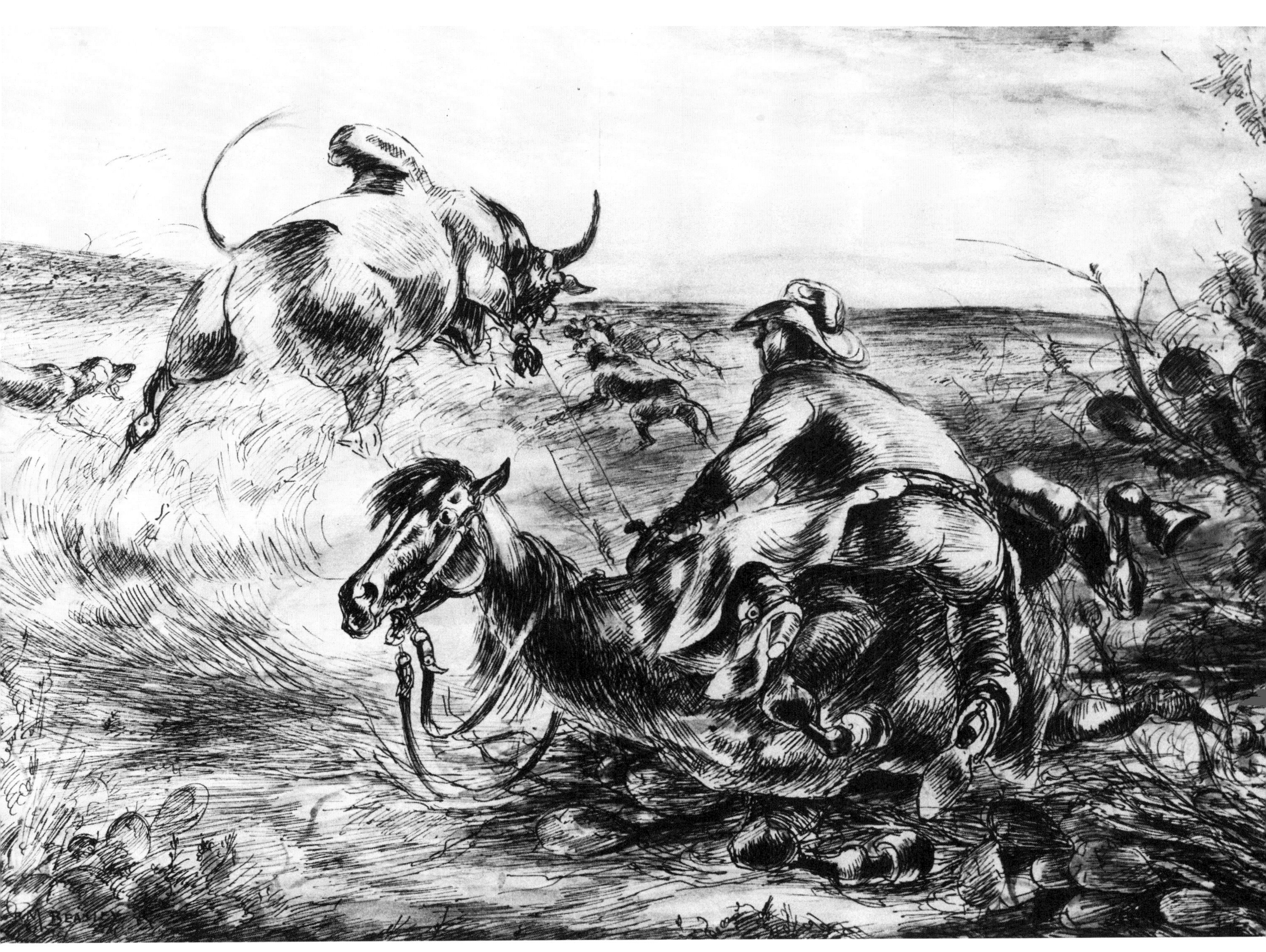

Strange is Destiny

Strange is Destiny, right where this Vaquero,
Rodolfo Muñoz happily told me this story
He was shot to death a few days later
He was happy because he had narrowly escaped death
When he roped a wild Brahman bull
That jerked his horse down as he jumped over a cactus
And drugged him for a piece.

The bull could have killed both of them
If it hadn't been that at that moment
The dogs caught up and crowded him.
It broke the rope and chase him away and save their lives.

I happened to arrive at the last of Rodolfo's agony.
He laid with his head toward the Southwest,
Tilted to the right.
And from the corner of his mouth
Blood dripping down slowly,
Like tolls of Requiem when a life is gone.

(El Coyote, *se llamaba el caballo* [the horse was named], by
    reasons unknown, he was the color of a coyote, I knew him.)
Ricardo Moreno Beasley

Death was the Author

Death was the author and the life of this story and the end of another, as that of a mother at the light of a child, who, innocent of his tragedy smiles with delight. Innocence is the glory of infancy, a misfortune at the age of reason, and a tragedy at the age when curtains are coming down.

It is in a hospital room where the story comes back to life, though by a tragic source, the murder of Rodolfo Muñoz and the wounding of his father. For Santiago, innocent still, of the cruel truth of the mourning that awaits his heart, to know that the same gun that downed him, killed his son at the same moment.

The mourning began as when the curtain rises to the drama of a tragic play, as if death closes one door to open another, when a visiting Vaquero, Sam Smithwick, with hat in hand, enters the scene of a humble hospital room. With reverence, he walks through the room and gives his hand in sympathy to all, in their mourning. As he looks down on his nearly dying friend, Santiago Muñoz, he pulls out his handkerchief to dry his tears, and with that, Sam Smithwick begins the story.

"I owe my life to this man," says he in a sobbing whisper to the friend that came with him, Willie Mancha.

"Once a wild steer was about to kill me, and Santiago saved my life by dashing his horse at the extreme speed into the horns of the steer, roping it at the same time, and jerking it to one side with his horse already bleeding in anguish and agony."

"Santiago was a very brave man, a fast thinker," Sam told me later. Brave he must have been, for at that late age, he did survive from the edge of the grave but not from the fatal truth—the murder of his son.

May this memory long buried in oblivion of a great past, live long in their honor.

> —Ricardo Moreno Beasley,
> February 9th, 1984

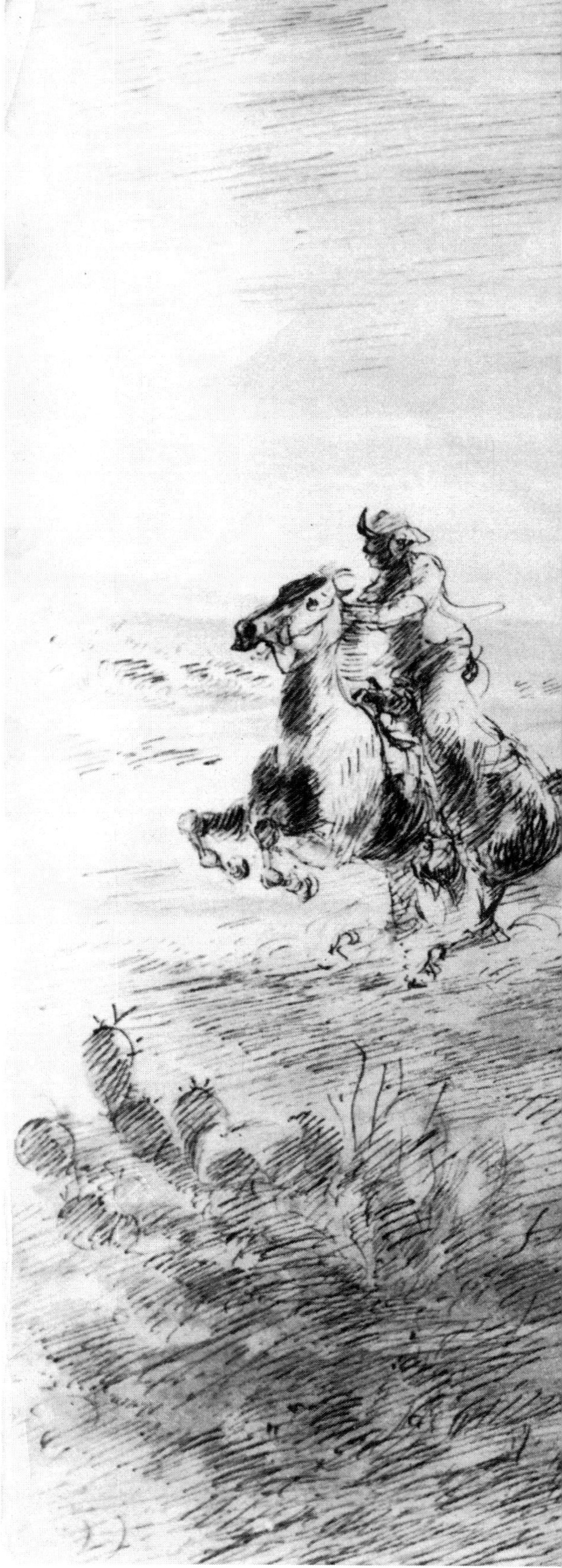

DRAWING 12
Dash the Vaqueros. © R.M. Beasley Legacy LLC.
Reprinted with permission.

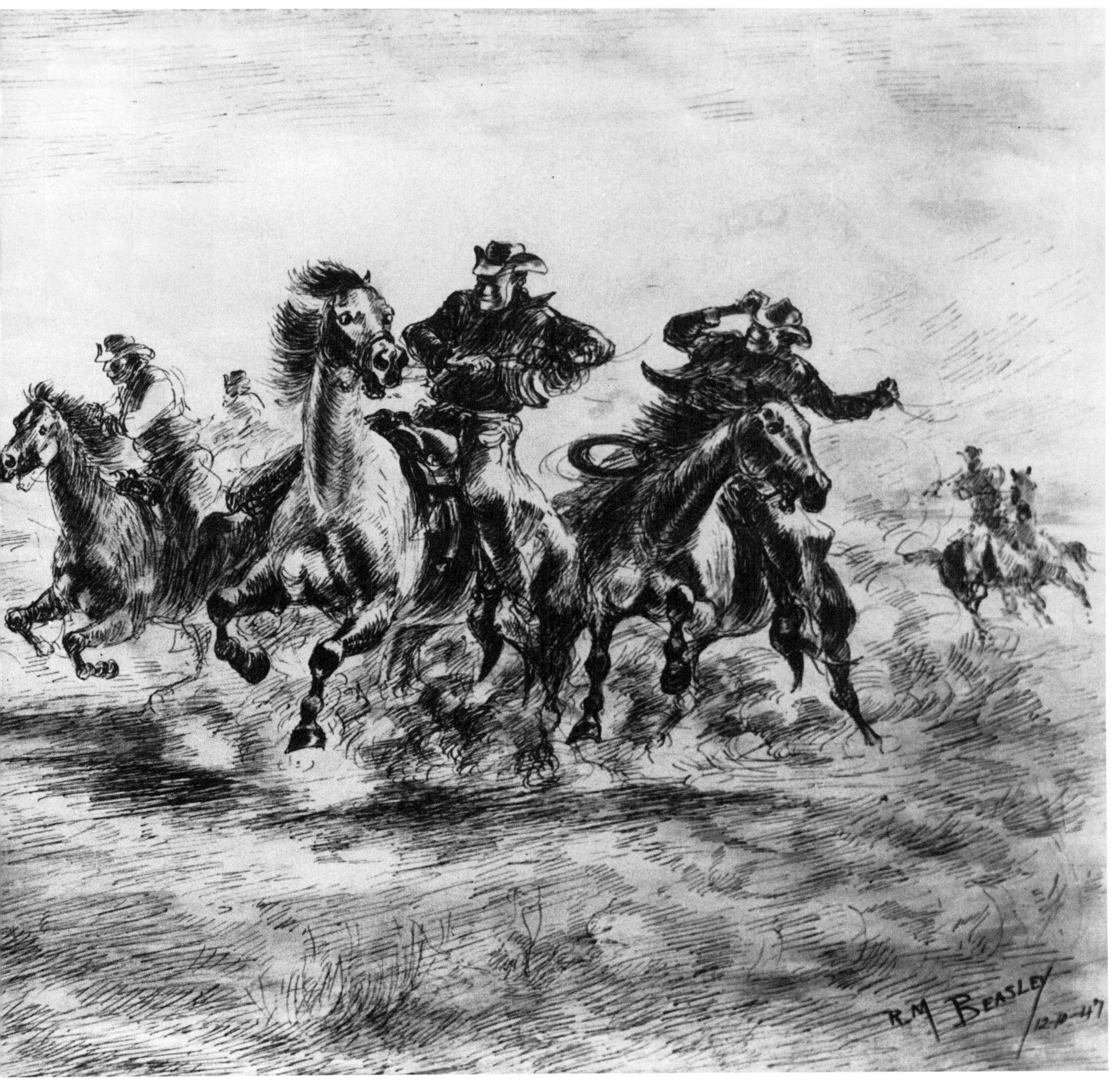
R.M. Beasley
12-10-47

Adios

To the end of the story
Of the life of the wild steer
Dash the Vaqueros.

As if for the last goodbye,
As if for the last sigh of a life.
Lamentable,
Unknown to them at that moment,
Was that ending and for time to come

Until they had to face the bitter revelation of the truth:
The era of the wild steer was forever over;
Never again,
Never more.

So unnoticed that parting had been,
The way it went away,
As that of the passing of a life in a sleep,
As that of time that never returns.

Should the Vaqueros have known that was the ending,
That unanticipated, unrecognized farewell
Would never have come.
The last of the wild steers on that range of La Gloria
Would have been let go to the end of life
As a tribute to its honor,
To let it come to its end and that of its time
In the dignity it deserved,
By the will of nature, and not by the hand of man.

The challenge presented

By the courage, intelligence, and determination

Of the wild steer to win its battle

Was the whetstone upon which were sharpened

The abilities of man and horse

To produce the best of the cowhorses,

The best of the Vaqueros,

Its glory.   R.M. Beasley

CHAPTER 2

The Duval Brush Country

Editor's Note

The ranching frontier that served as the setting for Beasley's vaqueros was born when the Spanish founding families claimed their land grants along and north of the Rio Grande between 1745 and 1755. About two thousand of these families had been recruited by the Spanish royal proprietor, José de Escandón, the Count of Sierra Gorda, who gave them large land grants signed by the king of Spain. These families, many of whom came directly from Spain, quickly intermarried with each other rather than with Native Americans, as happened in other parts of New Spain (the colonial name for Mexico). These families were the first of the Tejanos who raised and bred the six million distinctive Texas longhorn herds in what later became known as the cattle kingdom. These are the ranchers who established the legendary "rancho grande" and the accouterments of the American cowboy. They also bred the famous Texas mustang, known in Spanish as the mesteño. Like the Tejano longhorn, the mesteño has as a unique DNA and animal conformation. Both had their origins in the South Texas ranching frontier between 1750 and 1850. These families also bred oxen, mules, donkeys, and enough sheep to make Corpus Christi a world leader in wool export toward the end of the nineteenth century, when Anglo American ranching corporations and investors made Duval County their home.

Beasley noted that the Tejano ranching frontier stretched roughly from the Nueces River, along the Gulf Coast to the Rio Grande, and west to Laredo. It is distinquished from the other regions of Texas by its terrain, its climate, and its demography—primarily Tejano or Mexican American. Throughout most of the 1900s, South Texas had only a few large cities—Corpus Christi, Laredo, and Brownsville. It was mostly large ranches and rural towns. Duval County is in the geographic heart of this ranching frontier, and San Diego is the county seat.

The ranching country between the Rio Grande and the Nueces River is characterized by a rocky terrain with miles of dense scrub brush, which Tejanos called *chaparral*. Most of the shrubs that make up the chaparral were thorny or had thin foliage. It does have some areas of tall grasses and a few groves of small mesquite or oak trees, but most of the land offers little shade from the sun and elements. It is also characterized by a semiarid climate that varies from very hot most of the year to windy, dusty, and cold in winter. It was particularly vulnerable to the drought and dust storms in the mid-twentieth century, as described in Beasley's art and narrative. Beasley mentions in his narrative that Anglo American farmers immigrated to South Texas from the Midwest in the late nineteenth century and converted much of the Rio Grande plain into farmland. The central part around Duval County, however, has retained its rugged, ranch character and its Texas-Mexican demographic.

Duval County has many Anglo American surnames among its prominent families who came to identify as Tejano or Mexican American. These families, with names like Smithwick, Watson, Everett, Dix, Rogers, Stillman, and of course, Beasley, descend from military and political leaders who came from the Unites States and stayed in South Texas after the U.S.–Mexico War of 1848. Like Ricardo M. Beasley, they have a strong Anglo American or European Spanish appearance, but a deep pride in their Spanish and Mexican culture and heritage. Their heritage is their pride, and it is deeply rooted in the land, the ranches, and the vaquero identity. That heritage and the unique characteristics of these vaqueros inform Beasley's poems, his narrative, and his distinctive vernacular in the recorded interview transcripts. ★

Brush Country

Brush country, beloved land, home of the vaquero,
Of the wild horse, of the wild steer, as well as my own,
I will speak of you,
Mother of the Free.

Mother Nature made of you a paradise for the wild,
A heaven on earth and a hell to the vaquero and his horse.
Mother Land, you were great, you were kind and cruel at times.
But it was fair.

You made the vaquero great,
Because of you the vaquero excelled to the highest,
Because of you the steer got the wilder,
Greater were the dangers.

So much the better
The vaquero had to be to cope,
To survive through the verge of death,
To fulfill a duty.

Mother Land, you gave shelter,
You gave shield to the free.
You gave life, and to me, the cradle of my life,
The greatest inspiration.

You were the scenario of the drama
Of the life of the vaquero,
The goal of my life, for which
My greatest love will be for you
More than any other land.

*The brush country lies between the Nueces River and the
Rio Grande River in the south end of Texas, the Gulf of Mexico.
It was all a cow country by the beginning of the century, before the
farmers came in. It is made of three different regions.*

*The low land which runs along the coast, clear to the Rio Grande, was
a high mesquite country.*

*The sandy country which lies close to the Rio Grande and stretches out
from Hebbronville, east, clear to the coast. It is a lot of open country with
some heavy live oak stretches which was harder on the Vaquero and his
horse because of its toughness.*

*To the West is the hilly country, rolling hills, the brush is mostly Guajillo
and black brush—the chaparro negro. The Guajillo is a good cattle grazing
in some places. The hills get pretty rough with brush, but the worst of all
is the low places like along draw creek, river [sic] in some of the places all
this brush grows together, such as mesquite, black brush, cactus, all kinds
of raw thorny brush. It got to be such an infernal thicket, so cruel to man
and beast, that when the vaquero had to race through to rope some wild
steer, by the time he went through, his horse was bathed in blood from the
chest down and on both sides. It was pitiful to see such a sight, but the job
had to be done. It is admirable that as much the horse as the vaquero didn't
care the misery they had to go through to comply with their duty.*

I can tell you one description that I remember about the Brush Country.

Trail Drives

ANTES, HABÍA TRAIL DRIVES, como dice uno, que llevaban ganado aquí. Era terrreno abierto aquí; todavía no había entrado cercas aquí a Duval. No había cercas. Muy abierto.

Me platicaba a mi la señora de aquí de San Diego, Mrs. Southerland allá en noventa por hay, los 1880s, iban a Corpus en buggy, y era tan llano que tenían que llevar una estaca de fierro pa' persevar el caballo porque no había donde amarrar el caballo.

LONG AGO, THERE WERE TRAIL DRIVES, as one would say, that they drove cattle here. This was all open range; fencing had not yet come in here to Duval. There weren't any fences. Very open range.

That lady from here in San Diego used to tell me, Mrs. Southerland, around the nineties or so, the 1880s, they would go to Corpus in a buggy, and it was so barren that they had to take along iron stakes to secure the horses because there was nothing to tie a horse to.

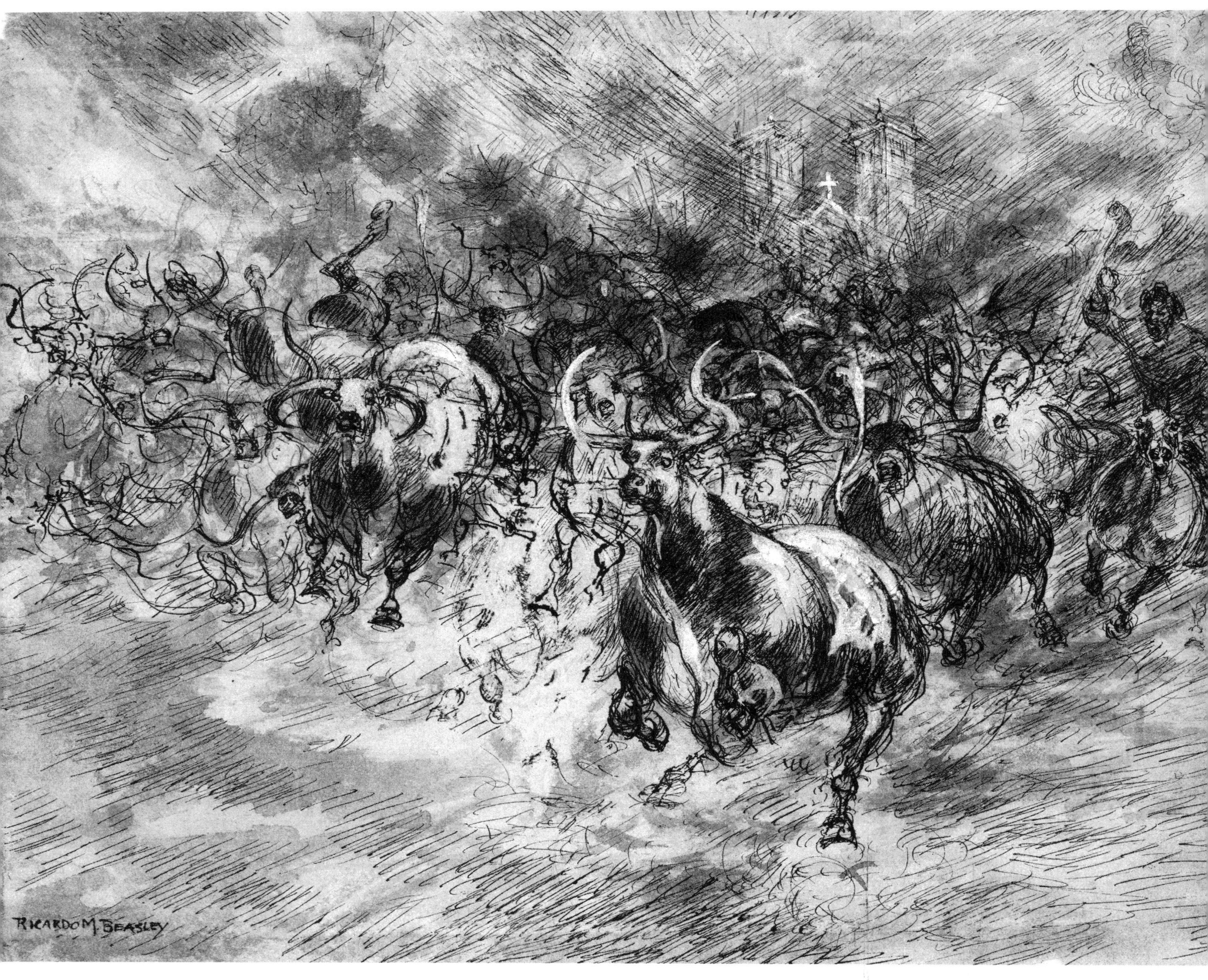

The San Diego Stampede of 1910

LET ME TELL YOU THE STORY about the estampida que pasó por la plaza de San Diego.

Llegaron Teófilo, el abuelo de Tommy Molina y los vaqueros de Manuel Rogers, que es el abuelo de Pat, de La Campana por esos novillos. Quinientos novillos aquí en San Diego. Y se los vendió a un tal Jim Luby, que viene siendo el padre o será el abuelo del mayor de Corpus. Y los iban a llevar para Petronila, un lugar que estaba cerca de Corpus. Era rancho de ellos allí, de los Luby.

La cuestión es que los llevaban para allá. Y le dice Manuel Rogers a Luby. Dijó, "No se vayan con los novillos por el pueblo porque vienen cansados y pueden dar estampida."

Un ganado cansado da estampida muy facil, bien nerviosos.

Dijó, "No se vayan por el pueblo porque pueden golpear gente."

Y no le hicieron caso. Se fueron por la calle de la iglesia. En ese tiempo pasaba el camino para Alice por la iglesia católica. Y luego volteaba para atrás. Allí llegando aca Angelita en la esquina había un puentecito de madera para el diche. El primero que pizó una tabla que hizo ruido, dieron estampida (drawing 13).

Todos.

Todos. Y se fueron por el corral de Angelita. Se llevaron la posta. Angelita me platicó que venía un hombre con niño en los brazos crusando. Porque en ese tiempo era plaza. No era parque. Corrió con él, y lo aventó arriba la cerca. Corrió y aventó. Ella lo pescó. Y él, yo no sé como se salvaría. Los vaqueros me platicaron que el ganado le dieron dos vueltas a la plaza. Como era plaza, había tablas puestas en bloque,

LET ME TELL YOU THE STORY about the stampede through the San Diego town plaza.

Teófilo, Tommy Molina's grandfather, and the vaqueros of Manuel Rogers, who is Pat's grandfather, all came from La Campana for those steers. Five hundred steers here, at San Diego. They had been sold by a certain Jim Luby, who happens to be, he must be the grandfather of the mayor of Corpus. Well, they were going to take them to Petronila, a place that was near Corpus. It was their ranch over there, of the Lubys.

The point is that they would take them there. So, Manuel Rogers tells Luby, "Don't take those steers through town because they've gotten tired, and they might stampede."

A tired herd stampedes very easily, very nervous.

He said, "Don't you go through town because they might injure people."

But they didn't pay attention to him. They went right down the street by the church. At that time, the road to Alice went right in front of the church. And then it made a sharp turn. Coming up to the street corner over at Angelita's, there used to be a little wooden bridge over the gully. The first steer that stepped on a wooden board and made a noise, ignited the stampede (drawing 13).

All of them.

Every one of them. They went right through Angelita's corral. They ripped out the corner post. Angelita told me that a man was crossing through, holding an infant in his arms. Angelita said she saw—he ran and he tossed it up. Back in those days, it was a plaza. It wasn't a park. He ran with the infant, and he tossed it up over the fence.

como asientos. En un ratito, no quedó una tabla. Que los novillos andaban arriba volando como un . . .

Y me acuerdo qué esto fué el año . . . en diez. En diez. Nineteen ten. You can see that church in the background, and those . . . es lo que me interesó para pintarlos, tu. Que con esa iglesia, se mira más violento. No sé si se entiende, pero para mi, le da . . . se mira una cosa brutal que pasara una estampida en una iglesia. Pero para mí, es un tema terrible. Porque si, en otras palabras, si no estuviera la iglesia allí, no se miraba tan violento como se mira.

Se puede ver movimiento y polvo y el vaquero. Déjame decir la historia de ese novillo.

No fué todo. Cuando andaban dando vuelta en la plaza . . . Manuel Rogers vivía allí en la esquina allí donde está todavía la hija de él . . . la Edna. Edna Rogers.

Trabajaba con Doctor De Hoyos. Edna vive en la esquina, casa de ladrillo que está. Y el corral de Don Manuel estaba donde está la iglesia, estaba el corral de él.

Manuel estaba ensillandola ya cuando la estampida se hizo. En la segunda vuelta que dió, se pasó el ganado derecho por la calle para el sur. Y allá le pescaba Manuel. Los hizo su punta y lo siguieron. Luego les dió vuelta la cuadra y luego los tiró por el arroyo—él y un tal gusto por hora. El arroyo iba medio cresido de agua, y quisieron llevar los novillos al agua—que eran quinientos—a ver si el golpe del agua los detenía un poco. Pero los novillos estaban más, más livianos y los cortaron.

Ellos no querían entrar al puente porque el puente era muy angostito, nomás como doce pies de ancho. El puente viejo de fierro que había aca por la calle de la Casa Blanca, la calle de la iglesia. Era un puente muy angostito hecho cuando no había carro.

Ellos le sacaron entrada allí por el peligro. Pero los novillo los cortaron y a fuerza los metieron adentro del puente. Y ya para salir del puente, dos caballos iban casi muchos.

Los alcanzó un novillo y los brincó por en medio y caió en tierra. Y como estaba llovido, se le fueron las patas y le caió atravesado a Juan Manuel adelante. Voló al caballo el novillo y siguieron corriendo como seis, siete millas para

Incredible. As for him, I don't know how he saved himself, but the cattle went around the plaza two times, the vaqueros told me. Since it was a plaza, there were benches made of boards laid over blocks. In a little while, not one board remained, because the steers were flying over like a . . .

I think this stampede happened in the year. . . in ten. In ten. Nineteen ten. You can see that church in the background, and those . . . that's what appealed to me about sketching them. That with that church in the background, it looks more violent. It gives it . . . it looks a brutal sight for something like that to occur at a church. Because for me, it's a theme, a terrible theme. Because if, in other words, without the church there, it wouldn't look as violent as it looks. And you can see the movement and the dust and the vaquero. Let me tell the story about that steer.

That was not all. While the cattle were circling in the plaza . . . Manuel Rogers lived right there on the corner where his daughter still lived there in the . . . Edna. Edna Rogers.

She worked for Doctor De Hoyos. Edna lives on the corner, a brick house still there. And Don Manuel's corral was where the church is now, that is where his corral was.

He was saddling up just at the time of the stampede, just when it happened, and at the second time around, the herd passed directly through the street to the south. And that's where Manuel caught them and gave them his lead and they followed him. After another turn around the block, he then led them down the gully. He and another hired hand. Now, the creek was half flooded. They tried to lead the steers into the water—there were about five hundred—hoping that the weight of the water would slow them a bit, but the steers were very, very volatile and the steers cut sharply.

The vaqueros didn't want to go on the bridge because the bridge was very narrow, only about twelve feet wide. The old iron bridge that was over here by the street that . . . by the Casa Blanca, down the church street. It was a very narrow bridge built before there were any cars.

They led them in there at their own risk. But they cut off the steers and forced them in. And by the time they crossed the bridge, two . . . well I have to say that on that bridge, two horses barely fit.

cuando se pararon. No iban dejando cerca. No iban dejando nada. Ya mero llegaban a Benavides.

Manuel Rogers era un hombre muy vaquero. Muy de a caballo. Ahorita, I'm working on that story, on that drawing. De donde brincó al novillo (drawing 14).

En este dibujo ya mero se lo llevaba a la fregada. Pero sabes que lo salvó porque se le ocurrió cada vez que venía el novillo a caerle arriba, le . . . La res no podía aguantar que le toque las narices. Es muy delicado. Cada vez que le venía el novillo a pisarlo, le tiraba golpes a la nariz. Y lo brincaba el novillo. Volvía a venir el novillo, le tiraba golpes, y lo volvía a brincar. Hasta que se enredó en las ramas, y se caió el novillo. Pero lo salvó que se le ocurrió pescarle las narices.

Y también como pasó que se tumbó su caballo. Ese caballo, pues lo lazó y se reventó el cincho y el novillo se devolvió. Y el caballo corrió. Vino el novillo, y le dió un golpe por detrás. Lo tumbó. Y de allí para allá no agarró a matar. Cuando el novillo no tiene cuacos, saca de pizar para matar.

This man lost his saddle. La reata se reventó y . . . nomás alcanza pegar el suelo. Es muy, muy interesante.

One steer caught up with them and jumped in between them and fell to the ground. Since it was rain soaked, its feet flew out from under it and it fell crossed in front of Juan Manuel. The steer flew over the horse and they kept running about six, seven miles before they stopped. They left no fences along the way. They left nothing. They ran almost to Benavides.

Yes, Manuel Rogers was a real cowboy of a man. Very good on horseback. Right now, I'm working on that story, on that drawing. Of the jumping of the steer. It looks like the drawing of that horse. In this drawing, it almost finished him off (drawing 14).

And what saved him was that he realized that each time the steer would come to fall on him, he . . . A cow can't stand to have his nose touched. It's very delicate. Every time it came to stomp him, he'd throw a punch at its nose, and the steer would jump over him. It would come back, he'd punch it, and it would jump over him again. Until the steer got tangled in the branches, and it fell down. But what saved him was that he remembered to catch its nostrils.

But how it happened that it threw his horse was that when he lassoed it. The saddle cinch broke and the steer came back. The horse ran off. The steer came and gave the horse a shove from behind, knocking it down. And from here to there it didn't get to—when a steer has no horns, it tries to stomp to kill. So this man lost his saddle, and you can see everything including the rope that broke and . . . it hardly reaches down to the ground. It's very, very interesting.

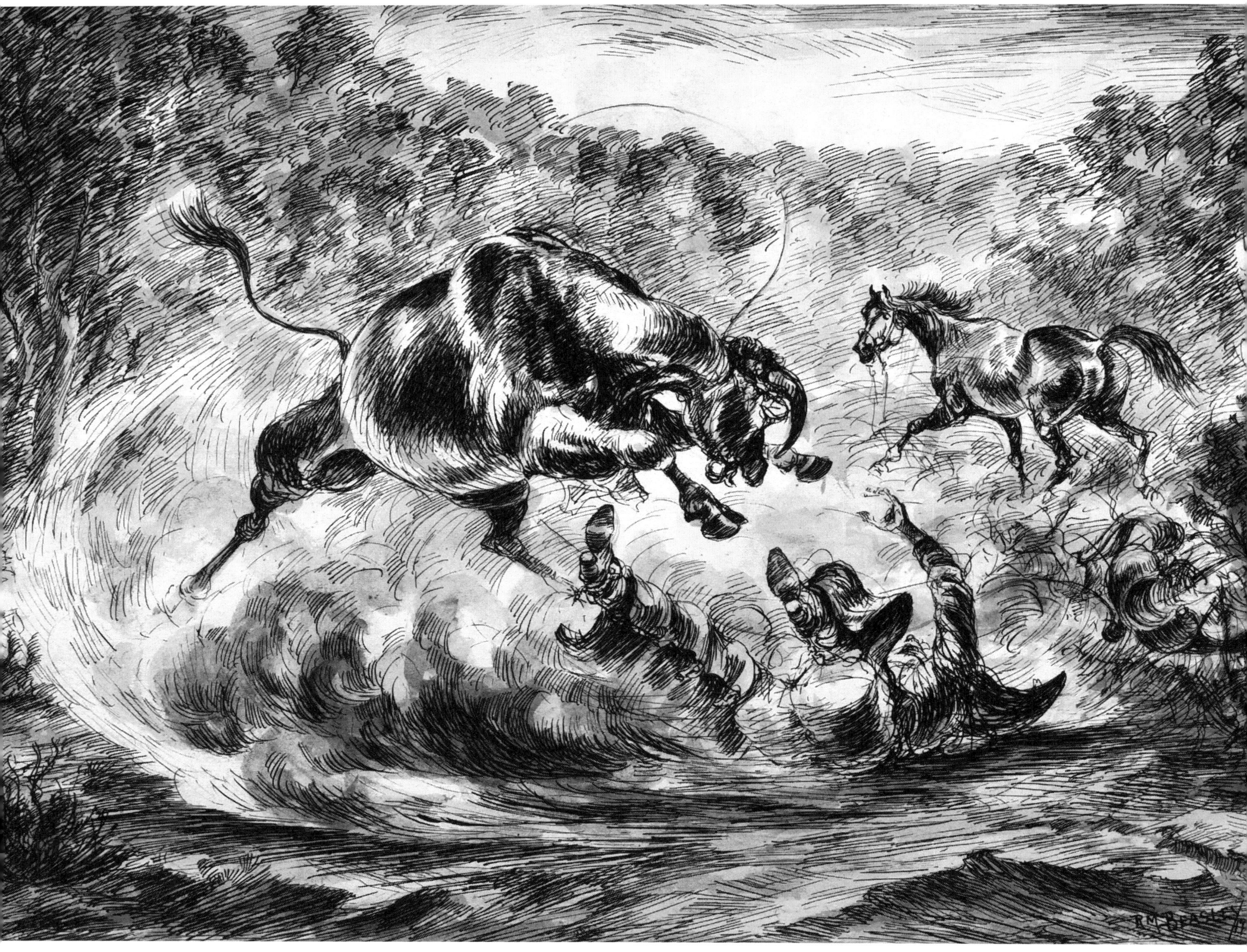

DRAWING 14
The steer came to stomp Don Manuel
Rogers. © R.M. Beasley Legacy LLC.
Reprinted with permission.

Drought

PERO LOS NOVILLOS ladinos ya se habían acabado aquí.

By forty-six. Yo vide allí cuando los pescaron el el arroyo vi estos novillos en La Gloria, y ya nunca volví a ver. En otras palabras, los rancheros dejaron de hechar novillos, porque al novillo lo criaban hasta que tuviera en veces ocho o dies años. Pero se acabó ese kind of business, y ya ahora los ves el becerro chiquito.

El puro becerro chico.

Ya para el cincuenta, ya no había novillos aquí, Duval. Pero en cuarenta y seis en La Gloria ya se acabaron porque yo andaba allí con ellos.

BUT THOSE DAYS end for the wild steers like these ended in the Brush Country.

From what I knew, here in Duval, in forty-six—by forty-six, it had ended.

I did see over there when they caught some wild steers in an arroyo. I did see those steers in La Gloria, and I never again saw them. In other words, ranchers just quit putting steers out, because that's how they raised steers until they were sometimes eight or ten years old. But that kind of business ended, and then you see only little calves.

Just little calf.

By the fifties, there were no more steers here in Duval. But by forty-six in La Gloria, there were gone because I was there with them.

DROUGHT

Drought, cruel monster that ravages the land like hungry fires
And only leaves the stench of death, despair and desolation,
Among human and beast.

The horror of starvation glitters in the scattered bones over the range.
How somber are those evenings of barren skies and dusty horizons,
When the sun goes down under the earth,
It has the semblance of a funeral in the silent of the night.

Every thing moves slow, pensive,
The hungry cows, the calves dragging behind,
Buzzards flying low in procession to land for the night near by,
As if in warning.

A man alone, coming home to an empty house;
The man who in vain scans the skies for a hope of rain;
The man who in vain scans the night horizon for a flash
Of the light of hope, of the flash of lightning.

Even the whirls, skeletons of dust, phantoms of the drought,
Move slowly, very slowly, four, five, six at a time,
All in different directions, a dangerous sign,
For when they come, they bring fear, the heart sinks at the sight,
It is like repeating to your ear the dreadful word,
Disaster, disaster, disaster,

And rain seems to be more distant than ever,
One gets the feeling that the drought will have no ending,
And we tend to believe it,
As tragedy has no bounds.
Ricardo Moreno Beasley

DRAWING 15
Drought. © R.M. Beasley
Legacy LLC. Reprinted
with permission.

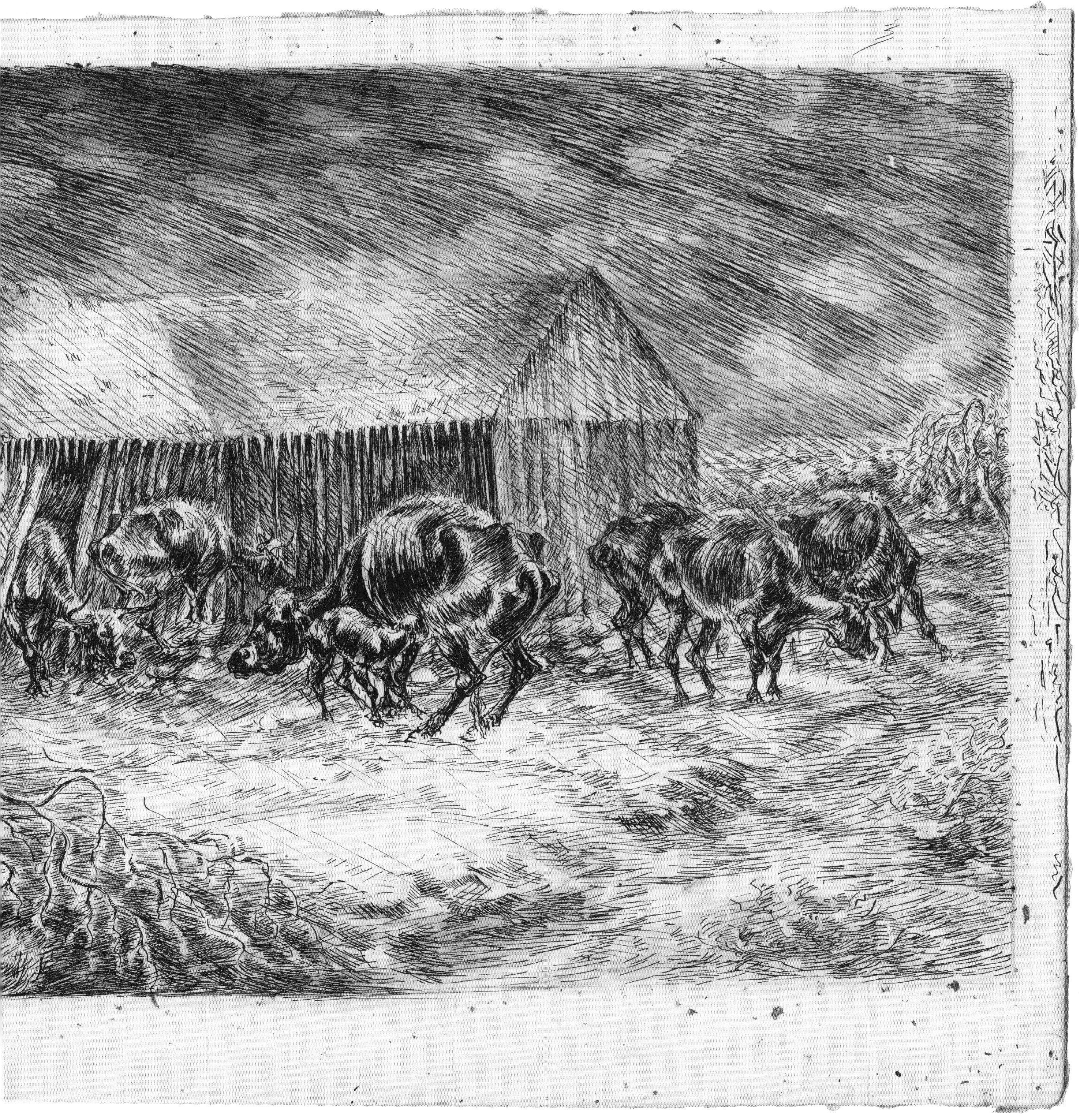

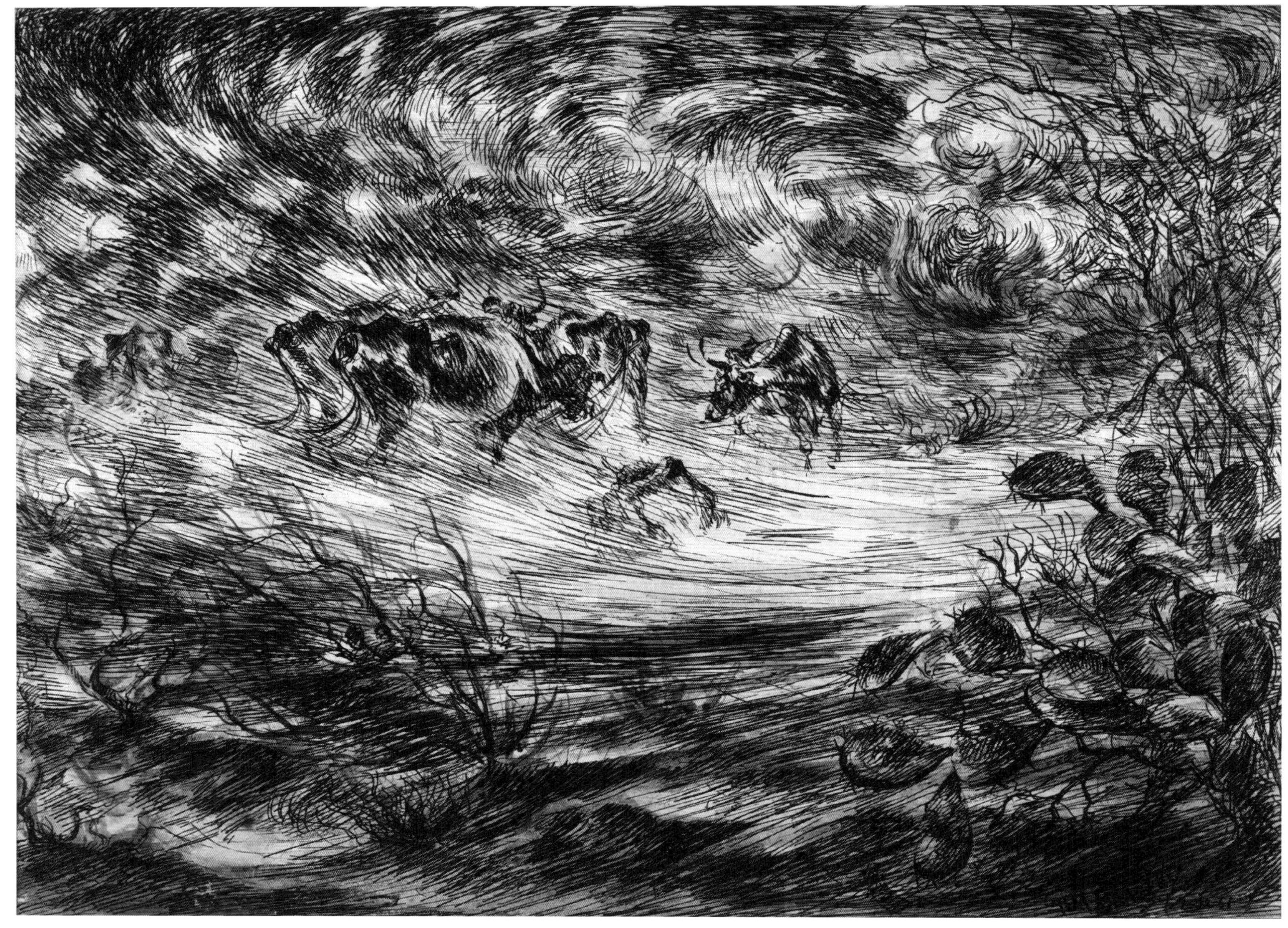

DUST STORM

Sometimes, in the dust storm, like in madness the,
Whirls go wild, tearing themselves to pieces.
Some will rise high up toward the sky, as If to God,
Imploring rain for the barren Mother Earth,
While others in agony, stagger on,
As the thirsty, hungry cows
That no longer know where to go.   R. M. Beasley

DRAWING 16
Dust storm. © R.M. Beasley Legacy LLC.
Reprinted with permission.

Dust Storm

TEMPESTA DE ARENA. Eso fué cuando empezó la seca de cincuenta. En cincuenta y uno, un domingo en la mañana me tocó a mí estar en el campo a esa hora de la madrugada, como a las cuatro, y llegó una tempesta arena del northwest. Y para cuando aclaró, ya no se miraba a cielo y tierra, y duró veinte y cuatro horas a esa fuerza hay. Lo que nunca se había visto aquí, esa clase de tempestad. Pero dejó la tierra tan barrilla como si hubieron sido dos años de seca [drawing 16].

Esta fué la seca que, de los siete años que dicen aquí. Aquí es cuando empezó.

Hay empezó. In the early fifties.

Primero venían las candelillas, de día se quemaba todo, y luego venía la tempestad de arena. Se acabó todo.

You can see the dust. You can feel it flying.

Qué horrible!

Y los ganaderos, los rancheros perdieron bastante durante esos años.

Casi todos vendimos, yo me tuve que salir. Muchos se salieron. ¿Qué, qué hacía ya? El potrero estaba pagando renta de oquis. No había ya por qué estar allí. Esas vacas, allí estaban junto el campo mio, dandose vuelta. Unos se hechaban. Ya se levantaban como los camellos del desierto, cansados de la vida. Quedó en la cerca así hundida de este alto de la tierra, amantonada. Pero yo nunca había visto antes una cosa de esas . . .

DUST STORM. That's when the dust storm of fifty started. In fifty-one, on a Sunday morning. It happened that I was the one out in the field at that time the dawn hours at about four o'clock, when a dust storm came in from the northwest. Before it cleared away, the land and sky were not visible, and it lasted twenty-four hours at that strength there. The likes of which had never been seen here, that kind of storm. But it left the ground so bare as if it had been two years without rain (drawing 16).

This was the drought they call the seven years drought around here. This is when it started.

In the early fifties.

First came the hot days, burning everything by day, then came the dust storm. Everything dried up.

You can see the dust. You can feel it flying.

Just horrible!

And the farmers, the ranchers lost a lot during those years.

Almost all of us sold. I had to get out. Many got out. What, what could I do? I was paying rent for the pasture for nothing. There was no reason to spend there. Those cows, they were there near my camp site, just going in circles. Some would hunker down. Then they'd stand up like camels in the desert, tired of life. Left on the fence buried this high, was the dirt piled up. But I'd never before seen anything like that . . .

Sick Cow

HICE UN CUADRO que un vaquero que ya no pudo. La vaca caida. Eso me pasó a mí. Fué en cuarenta y cinco [drawing 17).

Esa vaca flaca se caió. Ya tenía una semana de estarle llevando yo de comer y agua y hacía por levantar pero no podía levantarse. Y ese día tenía el pico en el suelo ya muy triste. Y trataba de levantarla yo, y pues no podía solo, yo.

Y ví para arriba y inconcientemente, ví tres hauras que andavan rodeandose. Y esas no se equivocan. Ya cuando ellas dan vuelta, se va ir muriendo.

Este se puede considerar un self-portrait porque fuí y traje el lapiz al campo y el cuaderno y I made a sketch y se murió a las dos horas.

Y el becerrito—la vaca era parte de brehma y el becerrito también. Esos, el brehmo es muy interesado. A él le importa muy poco de amor. Él quiere de comer.

Se perdió de repente.

Y pensé, "¿Pues qué se haría?"

Pues fuí y me lo hallé hechado con otro becerrito. Lo hizo compañero que tenía buena vaca. A los tres o cuatro meses, estaba flaca la vaca y él becerrito estaba gordo.

Ja ja!

Sele había robado toda la leche y no quedó.

Ja ja!

Y la pobre vaca se murió.

Se murió. Se murió. Nomás la acabé de dibujar, este, a una hora se murió. Pero los que me hizo hacer el dibujo fué cuando vi las hauras.

I DID A DRAWING of a vaquero who could go no more. The fallen cow. That happened to me. It was in forty-five (drawing 17).

That skinny cow fell. I had been taking it food and water for a week, and it tried to stand up, but it couldn't get up. And that day, it had its snout in the dirt, already, very sad. Well, I'd try to pick it up, but I couldn't do it alone, just me.

So without thinking, I looked up, and I saw three buzzards that were circling. And they make no mistake. When they circle, something is dying.

This one might be considered a self-portrait because I went to the camp and brought a pencil and the notebook, and I made a sketch and it died after two hours.

The little calf—the cow was part Brahman and the little calf too. Those, the Brahman is very interesting. Love is of no concern to it. It wants to eat.

Suddenly it was lost.

I thought, "Well, what could have happened to it?"

Well, I went and found it nesting down with another little calf. It became its little mate because it had a good mother cow. At about three or four months, the little calf was fat, and the mother cow was skinny.

Ha ha!

It had stolen all her milk and left nothing.

Ha ha!

And the poor mother cow died.

It died. It died. I had just finished drawing it, and, an hour later it died. But what prompted me to do the drawing was when I saw those buzzards.

Dije, "Esto es muerte, seguro ya." Cuando un animal está caido, una vaca está caida, viene todas los días una haura a la misma hora y da una vuelta y se va. ¿Entiendes? Da una vuelta y se va.

¿Pero cuando se quedan? Cuando se quedan, se va morir. Y eso eran tres ya, ya no hay vuelta. Pero digo yo el instinto que tiene el haura para venir a la misma hora. Da una vuelta y se va. La naturaleza tiene mucho misterio.

Mucho misterio.

I said, "This is death, for sure now." When an animal is down, a cow is down, each day at the same time a buzzard comes and circles once and leaves. Unbelievable. It circles once and leaves.

But when they stay? When they stay, it's going to die. And that was the third time, no more circles. But I have to say, the instinct of the buzzard, to come at the same time. Does one circle and it leaves. Nature has such mystery.

A lot of mystery.

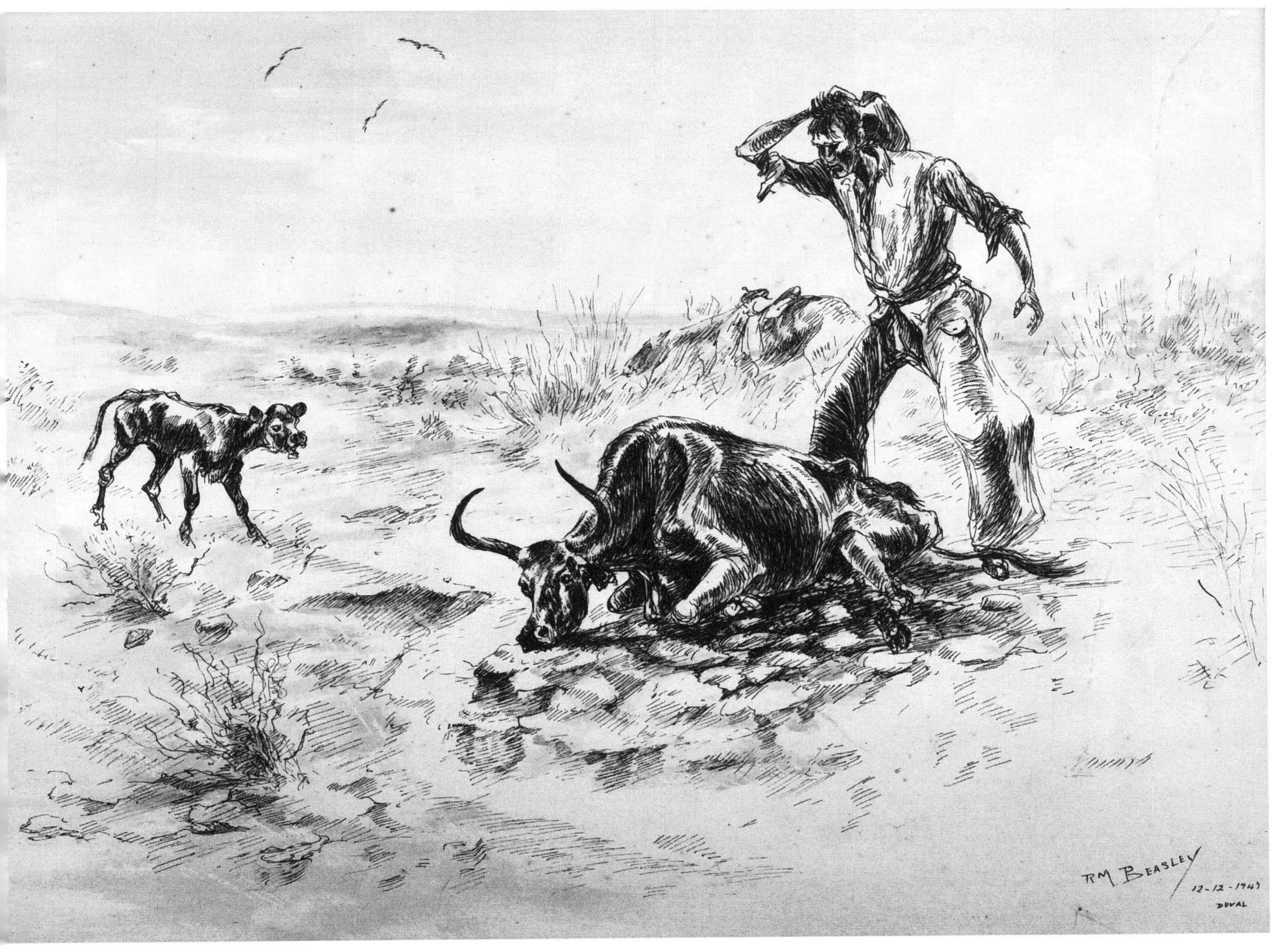

DRAWING 17
Self-portrait with fallen cow. © R.M. Beasley
Legacy LLC. Reprinted with permission.

El Chumascador

HICE UN DIBUJO del chamuscador. El chamuscador le quema las espinas al nopal para las vacas, que se comieran las ojas de nopal en tiempo de seca [drawing 18].

Este fué Juan Everett en la seca del cinquenta y uno, cuando era tempestad de arena.

Andaban chamuscando el nopal.

Chamuscando el nopal.

No. Más bién, andaba buscando a chamuscar, y no había nada. Y llega esa tarde y me dice al obscurecer, "Ya no hay que chamuscar aquí," Dijó, "Ya he andado todo el día y no he hallado una penca." Y Dijó, "Esas vacas se van a morir detrás de mí."

Y me impresiónó lo mortificado que venía. Y, I made a study of the drawing. Y él fué como lo viste, venía muy desesperado.

Y ni podía ver ni un nopal.

Nada.

Ya todo era muy . . . Si no nos vamos de . . De allí nos cambiamos para otro potrero, donde hallamos nopal.

I MADE A DRAWING about cactus singeing. A chamuscador is like a kerosene blow torch on a long pole. When the grass all dried up in a drought, the prickly pear cactus still had succulent cactus leaves with inch-long sharp spines. So the spines were singed off so the cattle could eat the cactus leaf (drawing 18).

This drawing was Juan Everett during the drought of fifty-one, in a dust storm.

They were singeing cactus. Chamuscando.

No, actually, they were looking for cactus to singe, and there was none. That's when he came up to me in the late afternoon, and just as it was getting dark. He says, "There's nothing to singe here." he said, "I've walked all day and I haven't found one cactus leaf." And he said, "Those cows are going to die right behind me."

I was so moved by how downcast he was that I made a study of the drawing. And he was as you just see him, very hopeless.

And he couldn't even find one cactus. Nothing.

By that time, it was very . . . If we don't leave from . . . From there we changed over to another pasture where we could find cactus.

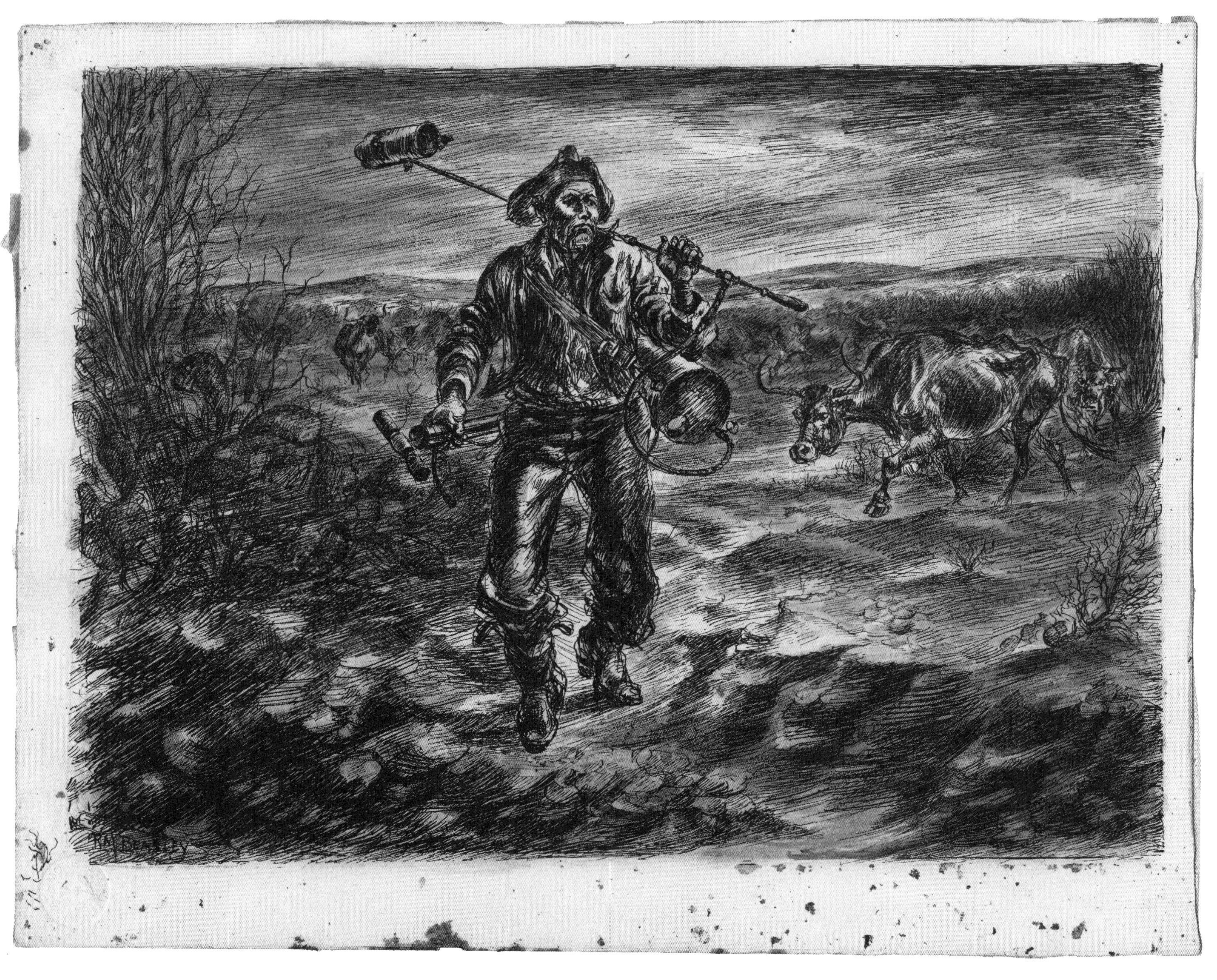

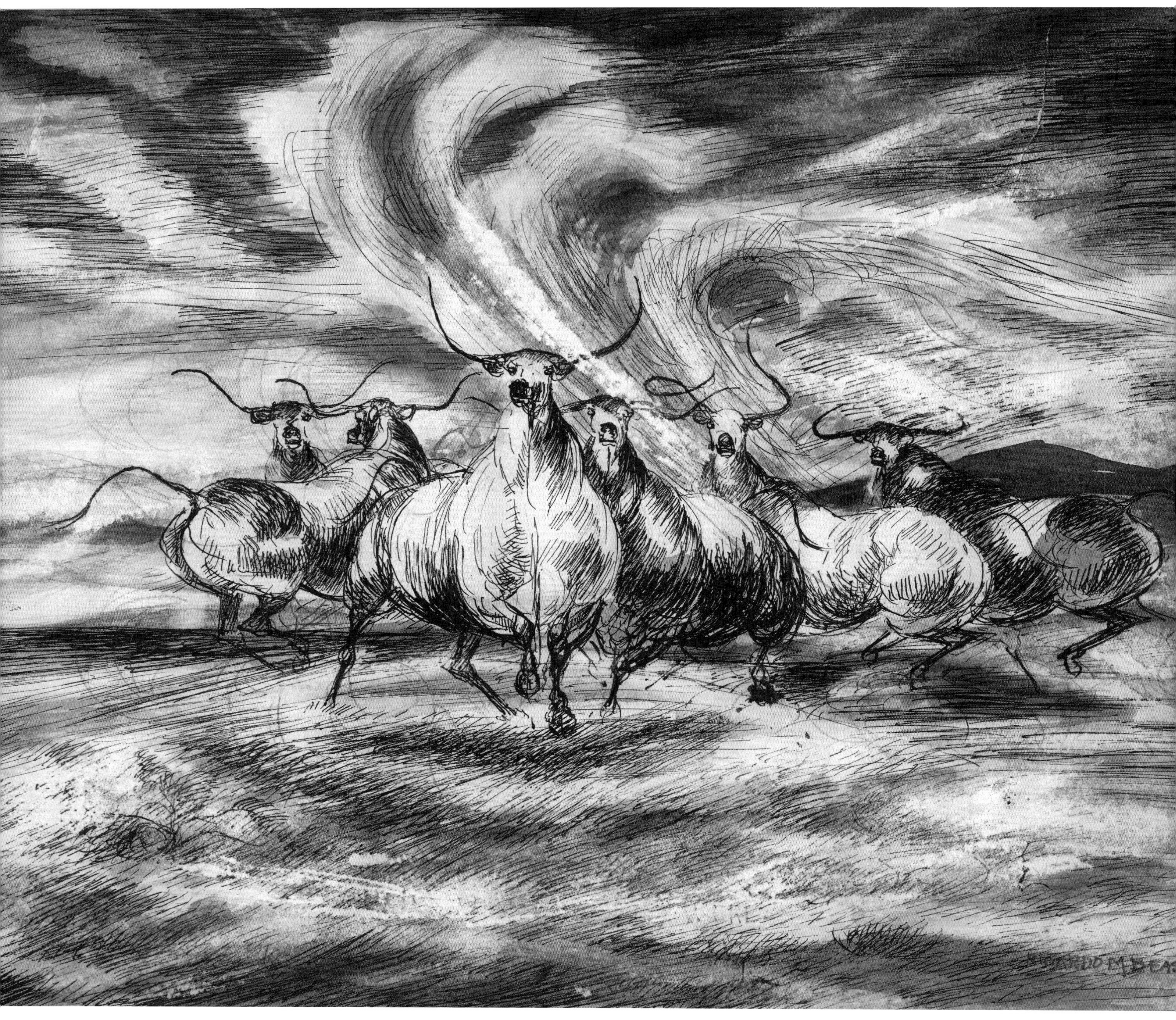

CHAPTER 3

The Wild Steers

Editor's Note

For Ricardo M. Beasley, a wild steer was the natural foil of the vaquero. They were natural enemies. The steer was born in the wild and had complete freedom on the open range. The vaquero roped the steer. He trapped the steer. The vaquero did not hate the steer, but he perfected the skills and tools to capture the animal. The wild steer hated the vaquero. Beasley's art was characterized by this natural conflict.

These were not the longhorns of the 1865 cattle drives. These were the descendant stock of the longhorn that had interbred with exotic stock like the Brahman. Indeed, the very term that Tejanos used for these animals, *novillo ladino*, meant interbred steer. *Ladino* was the term used for the mixed-parentage human offspring when the Spaniards first intermarried into their conquered Native American communities. Technically it meant *mestizo*, or mixed, but to the Tejanos, it had a connotation of wild or free-spirited.

The critical characteristic of a steer is that it is a castrated bull. The traditional method was for the vaqueros to round up the young male calves that had been born out on the open range and to bring them into pens to be branded and castrated. As steers, they were normally allowed to fatten for a few years before being rounded up and driven to market. But after the dissolution of the big corporate ranches around Duval County, these steers were allowed to grow much older, and to live out their lives in the mesquite thickets of the South Texas chaparral. Their interbreeding was combined with their isolation in the wild to produce a hearty, aggressive animal.

Beasley depicted the wild steers as cunning, demonic beasts. The vaqueros gave them names like Geronimo, El Mexicano, or El Zermeño—each with its unique behavior and character. El Mexicano, for example, crossed the Rio Grande back and forth, at will, from Mexico into South Texas. El Zermeño was an enormous beast that fought not so much for its freedom as its deeply felt sense of persecution, even revenge. The vaqueros and Beasley spoke of them as beings with feelings and emotions.

Although anthropomorphosis could be attributed to the quaint, uneducated perspective of the vaqueros on the remote ranges, the steers did have distinct mannerisms and behaviors that gave some degree of credence to the implied humanness. In one incident, Beasley describes the older steers as painstakenly avoiding the vaquero's constant traps. He wrote that they would physically prod the younger steers who blunder into the trap, and he gave the steers a voice to accompany the action. As a younger steer enters the trap, an older steer stops it with a jab in the ribs, as if to say, "Don't be a fool, there is danger about."

Beasley's steers are clever, defiant, although even the fiercest are captured at some point by the highly skilled and dauntless vaqueros. In the incessant struggle, however, Beasley invariably mourns the loss of freedom. Not the steer's freedom, but life's freedom. Nature's freedom. Indeed, in his art Beasley consistently illustrates even the captured steer as dominant. His own spirit is revealed in the domianance of nature over man and beast. Nowhere is this more evident than in Beasley's "El Mexicano" (drawing 21) and his poem "Revenge." ★

THE WILD STEERS

Like Phantoms, they were,
The remains of the Wild Steers
Against that horizon, where from beyond
They seem to reappear
On that moonlight of Duval.

Proud as ever,
Defying the Vaqueros
From whom they had vanished
For what seemed ages.

So mysterious they had become
As if they had never existed,
And like Princes of fairy tales
Had spirited away
Without a trace of their domain.
R.M. Beasley

(© R.M. Beasley Legacy LLC.
Reprinted with permission.)

IT WAS A MEMORABLE OCCASION one night my father and I were sitting at a table when I was a child of seven years. He just reached down and got hold of a paper and a pencil and said, "I am going to draw for you a vaquero roping a novillo ladino (a wild steer).
I watched him intently as he drew. When he was through, he showed it to me. I liked it very much. I was excited with joy. It was the first picture of a wild steer that I had ever seen. Although the drawing was crude, to me at that age, it looked great, majestic to me because I looked at it as something real—a real novillo ladino. It so inspired me, that I made it my life, my destiny, to follow the story of the wild steer, to the end of the trail, in his desperate battle with the vaquero, to be free.

A steer is a bull calf that was castrated as a calf. In the times of the big steer, when it was the leading market, ranchmen would let them grow to four, five, or even ten years old. In the big ranch country with the dense brush, these steers had the opportunity to grow wild, to be free.

The wild steer was a very intelligent animal. Since it had no sex, no urge of nature to lure him into danger in the search of a mate, it dedicated all its senses for its own protection; to take care of its freedom which was its most valuable possession.

The wild steers were brave, very proud of themselves. They went to the extreme of anguish, of thirst and hunger rather than to fall into a trap and lose their freedom.

Juan Everett told me that the wild steer could live on prickly pear for water and food a long time. They would hide out in some remote area of infernal thicket of brush and prickly pear and not move around much to avoid making tracks specially when they felt the presence of vaqueros around the country. They had sense enough to know that tracks are their worst weakness. Tracks don't keep secrets; they are written words on the ground that tell their whereabouts.

The extreme wild steer travels alone and very carefully, very slow. Only after he senses no presence of vaqueros for several days does he move toward some water hole, but very carefully, *de mogote en mogote*, from thicket to thicket.

He might spend days on each one just to see what he can hear, what are the telltale sounds around him. When he finally reaches the water hole, and if it is a windmill—and—the water trough is usually inside of a corral—he will be very suspicious to approach. He will watch from a distance, observe everything, every move of any other animals, cattle—even birds—to get any warning of danger as he well knows that the corral is a trap. He would even circle around it to get the wind, his best friend, to warning of a hidden danger.

To play safe, he would go away without drinking water, to come back the next night which is usually the time they choose to travel. It might take him hours to travel the distance of forty or sixty yards from the edge of the brush to the corral gate, if he so decided. Then he will check the gate, and the post where the gate closes with his best adviser—his nose—his scent, to get the warning if a human has recently touched it. If there is the least suspicion of danger, the presence of a vaquero around, he will sharply bolt, and storm away. But not from fear. They attacked, too.

Waiting for the Wild Steers

ESTOS NOVILLOS son de . . . La Gloria. Esos novillos eran tan ladinos que no pintaban huella, ni sabían que estaban allí. Que se espían.

Y un día, hallaron huellas en una loma donde se arrimaron cercas de un aguaje. Y decidieron los vaqueros a espiarlos de noche para lazar. Cuando se arrimaban, venían cerca, y los vaqueros llegaban al lugar a donde los espiaban (drawing 20).

Los novillos llegaban de temprano en un ramadero de un monte. Y los novillos no llegaban hasta la una o las dos de la mañana. Los vaqueros se acostaban a dormir. Pero los novillos, como venían en loma, hacían ruido.

Y los caballos eran muy inteligentes. Les gustaba mucho el excitement de lazar. Le volaban. Muy inteligentes. Todo el tiempo estaba uno o el otro en alert, escuchando. Uno dormía y otro velaba. Ya cuando sentían los novillos que venían, they would pull at the vaquero. El vaquero se amarraba la rienda en la mano, en la rodilla. Lo dispertaba el caballo. Pero los novillos nomás llegaban hasta cierta distancia, y luego ya empesaban a darse vuelta para irse.

Y ya para cuando llegaba el excitement a ese punto, mandaban a un vaquero que fuera a la orilla del monte a ver para la loma, a ver si los novillos . . . ya cuando se van a ir para atrás, empiezan a dar vuelta, dar vuelta, dar vuelta. Y si se adelanta uno, este, va uno de los novillos viejos y lo alcanza, y le da un cuacazo en las costillas y le dice, "Wake up." Para decirle que, "No seas bruto que hay peligro." Ellos también . . . keep them together.

Iba y lo cornaba y volteaba para atrás. Ya cuando estaban en ese punto, entonces sí salían los vaqueros—a todo vuelo.

SO, THESE STEERS belong to . . . La Gloria. Those steers were so wild that they left no track marks. Nobody knew they were there. That they were ever seen.

Well one day, their tracks were seen on a hillside where they had come up close to a watering hole. So the vaqueros decided to watch for them at night to rope them. When they came in close, the vaqueros went up to the place where they wanted to watch for them (drawing 20).

The steers came up early to a thicket in the brush. But they wouldn't come out until about one or two in the morning. The vaqueros would lie down to sleep. But the steers, as they came up the hillside, made some noise.

Well, the horses were very intelligent, and they just loved the excitement of lassoing. They would fly at it. They were so intelligent, one or the other was always on alert—listening. One would sleep and the other stood vigilant. And when they sensed the steers, that the steers were approaching, they would pull at the vaquero. The vaquero would tie the reins on his hand, or his knee. The horse wakes him up. But the steers would only come up to a certain point, and they would start to circle back and leave.

So, whenever all the excitement reached that point, the vaqueros would send one vaquero to the edge of the brush to look over at the hillside, to see if the steers . . . and just when the steers are about to circle back, they start turning back. Turning, turning. And if one of the steers starts advancing too close, well, one of the older steers catches up to it, and gives him a jab with his horns in the ribs like, "Wake up." To tell him, "Don't be a fool, there is danger about." They also . . . keep them together.

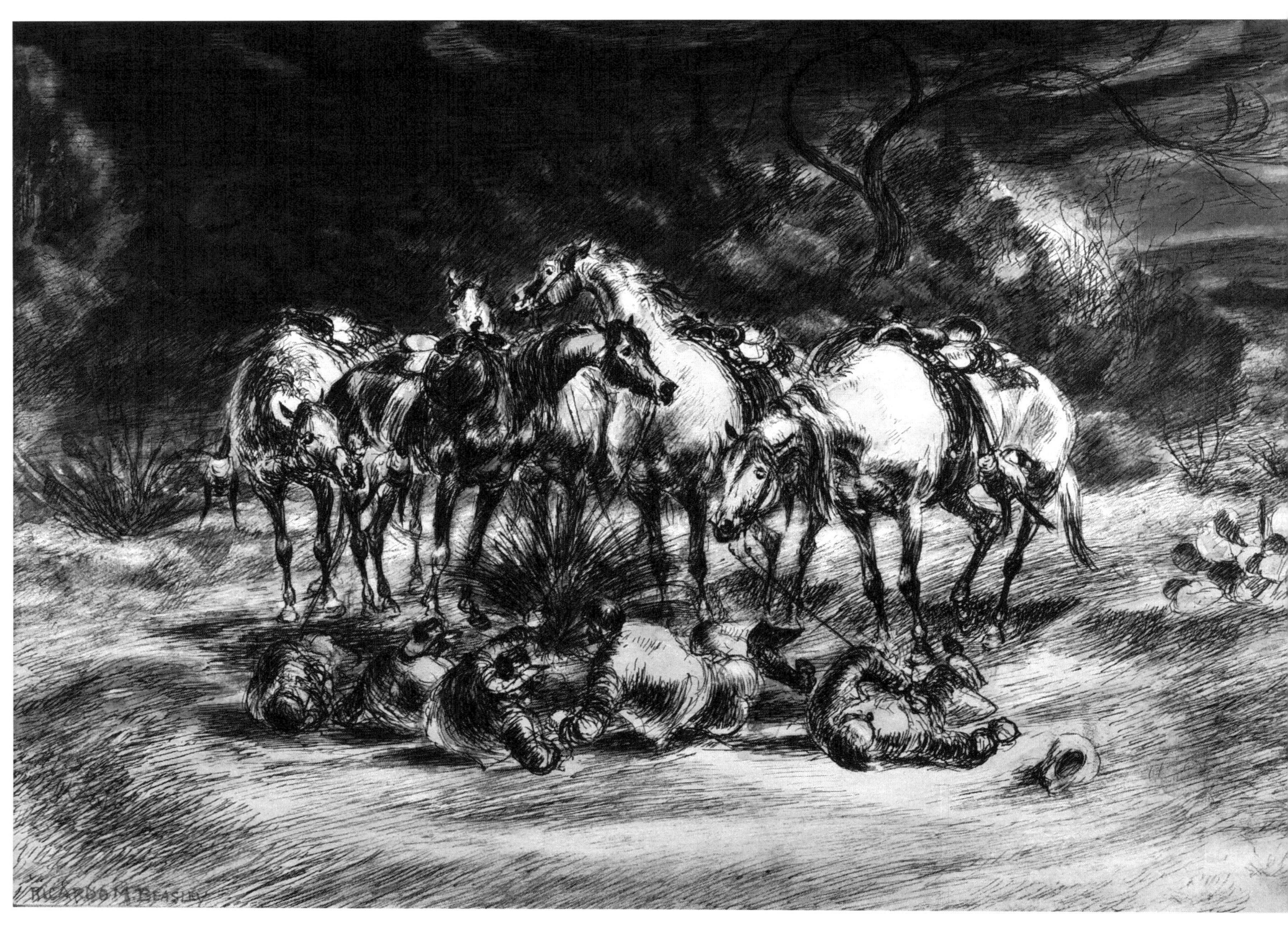

DRAWING 20
Waiting for the wild steers. © R.M. Beasley
Legacy LLC. Reprinted with permission.

Todo vuelo.

Y cuando la luna se apagaba, que se ponía una nube que se ponía obscuro de atiro, en veces, pues ya no miraban animal. Y luego el caballo parecía que iba corriendo para atrás otra vez.

Pensaba el vaquero, "¿Pues qué haría con este animal?" Y no sabías lo qué era hasta que te pegaba—iba avanzando hasta que te pegaba el polvo del novillo que te iba aventar por detras en la noche. El caballo solo buscando . . . Por el scent, por el scent, el smell del polvo. Había veces que caía en un arroyo el novillo, y caía el vaquero arriba con todo y caballo.

Ahora en estos días modernos que hacen mucha propaganda de the trained cutting horses y todo eso, y los vaqueros tenían eso mismo. Ellos mismos los entrenaban.

Ellos los entrenaban.

Y cuando era en un chaparral, necesitaban unos caballos que sabían . . . eran caballos del monte que eran . . . ¡Oh, qué bárbaro! Muy inteligentes. ¡Qué bárbaro!

It would go and jab him with its horns, and he'd turn back. But just as they would reach that point, well that's when the vaqueros would come out after them—at full speed.

Full speed.

And sometimes when the, the moonlight goes out, just when a cloud covers it and it gets completely dark, well, they couldn't see the cows. And then, in the dark it would seem like the horse was running back the wrong way.

A vaquero might think, "Well what is going on with this horse?" And you wouldn't know what was going on until you felt—the horse kept advancing, running until you got hit with the dust from the steer that was charging in the dark to hit you from behind. The horse, himself, was searching . . .By scent. by scent, the smell of the dust. There were cases when the steer would fall into a gully, and the vaquero would fall in on top of it, horse and all.

Nowadays that they make a big issue about the trained cutting horses and all that, and the vaqueros had that same skill. They themselves trained their own horses.

They trained them.

And in the brush, they needed those horses that were trained . . . Oh, amazing! Very intelligent. Amazing!

WAITING FOR THE WILD STEERS

In the nights of the moonlight of long ago.
When they were roping wild steers in the night,
The great spirit of the horses and the Vaquero,
Blazed more than ever had or ever will again.

So intense was the passion of that life
That it was like a romance with agony,
To court even death, so as not to fail,
As if something of a sacred duty.
—R.M. Beasley

I WAS TOLD OF A VAQUERO, Andrés Villarreal, to what extreme he had to go to avoid detection of his scent when lying in wait for wild steer. He would dig a hole big enough to get into and far enough from the gate to play safe to pull a rope that would shut it, should the steer go in. Then he would take his clothes off and smear his body with mud to avoid giving away his scent.

The wild steer would not come to water at a windmill unless all the water holes were dry, and even at that, some never did come. He would rather die or survive on the juice of the pear.

Juan Lichtenberger says that during the season of pear apple, the steer doesn't suffer from thirst because he eats it very much. He gets up very early in the morning to have his breakfast at the same time he takes advantage of the dew that is on the grass. Then he gets out to graze at the edge of the hills. If he scents or hears the approach of vaqueros, he dashes to the brush like lightening, and is gone for good with plenty of time before they arrive.

It is almost unbelievable the keen sense of observation the wild steer would develop for is safety. If a real close friend of mine, Teófilo Molina, wouldn't have told me of this observation, I am going to tell you about, I would have doubted that that much intelligence existed in a steer.

The wild steer, says Teófilo, considered the horse his enemy. Whenever he came across the track of a horse on his trail, he would get very suspicious and smell the limbs of brush on both sides of the trail for the smell of sweat that the horse may have rubbed on it; and if such was the case, he would change his course right away, but he knew right then that the horse was carrying along his worst enemy, the vaquero, because unridden horses don't sweat.

The wind is the best friend of the wild steer, his keen sense of smell, the wind keeps no secrets, it reveals what the eye cannot see, what the ear cannot hear, what the night conceals. The wild steer keeps his nose up to the wind most of the time for a message of warning of the approach of danger. His keen ear is his second weapon of defense.

Teófilo Molina told me about the best demonstration of the fear vaqueros had to be detected by steers. He said that when he and another vaquero had to lie in wait for wild steers at a certain windmill, they would do it on top of the platform of the windmill with a wire attached to the gate to close it to trap the steer. They had to be that high was to avoid spreading their scent.

First, when they got to the hiding place, they would take the precaution to leave their horses half a mile or a full mile away and unsaddle. When thirsty wild steers got wise about vaqueros lying in wait for them at any water hole, before approaching, they first would circle around it a ways off, to have the wind give the warning of their presence. The reason the vaqueros took the saddle off the horses was because the saddles would shake and make rattling noises that the steers could hear that a long ways off. That would be warning enough for them to fly away.

Another way that the wild steer shows real humanlike intelligence is when they have to cross open range. He knows there is a danger of being caught in it. But in some instances, if it is possible, he would take advantage of some gentle cattle going to water and follow behind them at a distance to make an easy get away. If there is any danger ahead, the ones in the front will give the warning. The steer plays it safe, and gets away before he is even seen.

Another example of his humanlike intelligence is his ability to hide in the brush when being chased. Suddenly he disappears, and it is not surprising to find him laying flat on his side as if he were dead, to avoid detection.

Some wild steers showed a sense of brotherhood among themselves. They seemed to care for the safety of the freedom of each other.

Teófilo Salinas told me of observing that act several times at a place near a windmill. Way deep in the night, a few wild steers would gather on top of a hill nearby. They would not come down to water. They would walk closer, but staying at a safe distance. And if one of them was lured by the smell of water and the anguish of thirst, and dared to

advance forward beyond the point of danger, then, one of the older
ones would trot ahead of him and hook a horn on his shoulder, and
turn him backward, like saying, "You fool, you are risking your freedom, your life." At that, they would circle around a few more times,
and leave the area. That's when the vaqueros would then dash out
after them to rope them by the light of the moon.

Seems unbelievable that the sense of loyalty would exist among
some of the wild steers to the extreme of endangering their own life.

Teófilo once told me that when wild steers ran in bunches, and
the vaqueros would rope one of them and tie it to a tree, some of its
friends would come back at night to visit with it. So loving were they,
that they would carress it by licking its shoulders. That was their
demonstration of consolation.

Sometimes they would stay with that steer until daylight because
that was when they ran the danger of getting caught as on the way
back to their hide out.

The Silent Cowbell

IN ANOTHER WAY the wild steer displayed his intelligence to the extreme was when wearing a bell. When the vaquero would hang a bell on the steer because he was hard to find, the steer, however, used his wisdom to his advantage to keep the bell silent. To the steer the bell was a treacherous dangle that betrayed his every step of his whereabouts. A constant spy that followed him like a dark shadow, sealing his fate with a noisy clatter.

Teófilo Salinas told me of a certain wild black mulley steer, from the Eagle Hill Country, that wore a bell with a chain. He loved his liberty as much as those patriots of 1776 that rang their bell to declare Liberty. To the black steer, ringing this bell could be the end of his own liberty. To keep that bell silent, the steer used all his wisdom. He snuffed it down by tightening his jaws against it with his throat to such an extreme that the bell couldn't ever whisper as he traveled along very slow and as cautious as if his next step was the trap of death.

Teófilo remembered of one night when they were laying for wild steers by a water tank. Suddenly the wild clatter of a bell stunned them as the black steer started off, so close, almost on top of them. At that instant they were in hot pursuit because their horses loved the thrill of the chase as much as the vaqueros did.

They were lucky. The steer in his excitment took to open country and that gave them the opportunity for a hot race, but it wasn't fast enough. The black steer beat them to the next thicket beside a fence line, and that was the last they heard of the bell. It went into a dead silence as if the steer had dropped dead.

It was kind of embarrasing for the vaqueros; the way the steer had just fooled them and still managed to make a get away. So to get even, the caporal (foreman), Feliberto Garza, decided he was not going to let the black steer get away with this humiliation.

They were going to set vigil all night long. And just to play it safe, he placed his men on sentinel all around the thicket so the steer couldn't have a chance to sneak away in the silence of the night with the silence of the bell.

All through the night, says Teófilo, they sat on their horses in agony with their ears open, intently listening for the least whisper of the bell. They wanted to make it sure that the master of the voice of the bell wouldn't have the least chance to get away.

By day light, not a note was heard of the cursed bell, so they thought for sure they had him trapped. As soon as it was clear enough, the caporal sent a man into the thicket to scare him out into the open country which was all around.

The vaqueros had it all figured out, there was no way in the world the steer could get away this time. As soon as he would hit the open country, one of them would rope him. So, with horses rearing to go, set for action and rope in hand, they waited for the final moment of a long story, almost any instant.

To their utter dismay, after a long wait, the man in search of the steer came out of the thicket disappointed with the bad news that the steer had gotten away. He had looked for him almost behind every bush and it just wasn't there.

"Oh, yes. He is there," was the answer of the vaqueros. "He couldn't have gotten away."

"Well he sure is not there because in such a small thicket, I couldn't have missed him as much as I searched," he answered.

"Well, you did," said Teófilo, "He is down there, all right. In a hole. You go back and right by the fence is a deep section of the creek covered up with green vines. I think that is the only place he could find to hide to his liking."

The man went back into the thicket and soon the wild clatter of the bell was heard as the steer shot out from the hole, running all out for his life. But the moment he hit open ground, his luck ran out. He soon was made a prisoner. He was roped and tied down. But not forever.

In the next few days, the black steer was sold along with some other steers to La Esperanza Ranch, the neighbor to the south—still wearing his bell. Martin González, who was the caporal at La Esperanza, told me that for a long time, they didn't see the steer again or

hear of the bell either. They had come to the conclusion that he had
passed out of existence or was roaming far away in some unknown
country as no sound was heard of him.

To their surprise, one day when riding through that country,
suddenly, the black steer jumped out of a thicket almost under their
noses. Like Comanches on the warpath, the vaqueros all tore after
him with wild yells. After a hot chase, they roped him and tied him to
a tree.

Happy over the conquest, Martin told the vaqueros, "This time
the steer will never get away. We will neck him to an ox and take him
to camp. We will keep him in the little pasture, and as soon as we are
through with the work, we will butcher him. So when the time came
one evening for the feast, Martin ordered his men to go and get the ox
with the steer tied to him by the neck. Tie him to a tree and take his
bell off, and turn the ox aloose, which they did. The butchering was
set to be done in the morning.

When they were tying the steer to the tree with the same rope he
had been necked to the ox, someone warned them not to use the same
neck rope because he was going to pull out his head of the necktie
and get away. He warned that the tie was too loose, too big, but they
all argued, "How can he ever pull his head out of the tie now, when he
has been necked to the ox for two weeks and never got loose?"

"Because," said the man, "All this time, the steer has been pulling
forward and the tie pulled back on his neck as he tried to push away
from the ox. Tonight, he will be pulling back to get away from the tree
and I am afraid the tie will slip out of his head." And that was all that
was said.

The next morning, which they thought was going to be a happy
day in camp, turned out to be a great disappointment, and a happy
one for the black steer. He had got away and without the bell to worry
about. He felt a lot more safe to remain free.

He was never seen again.

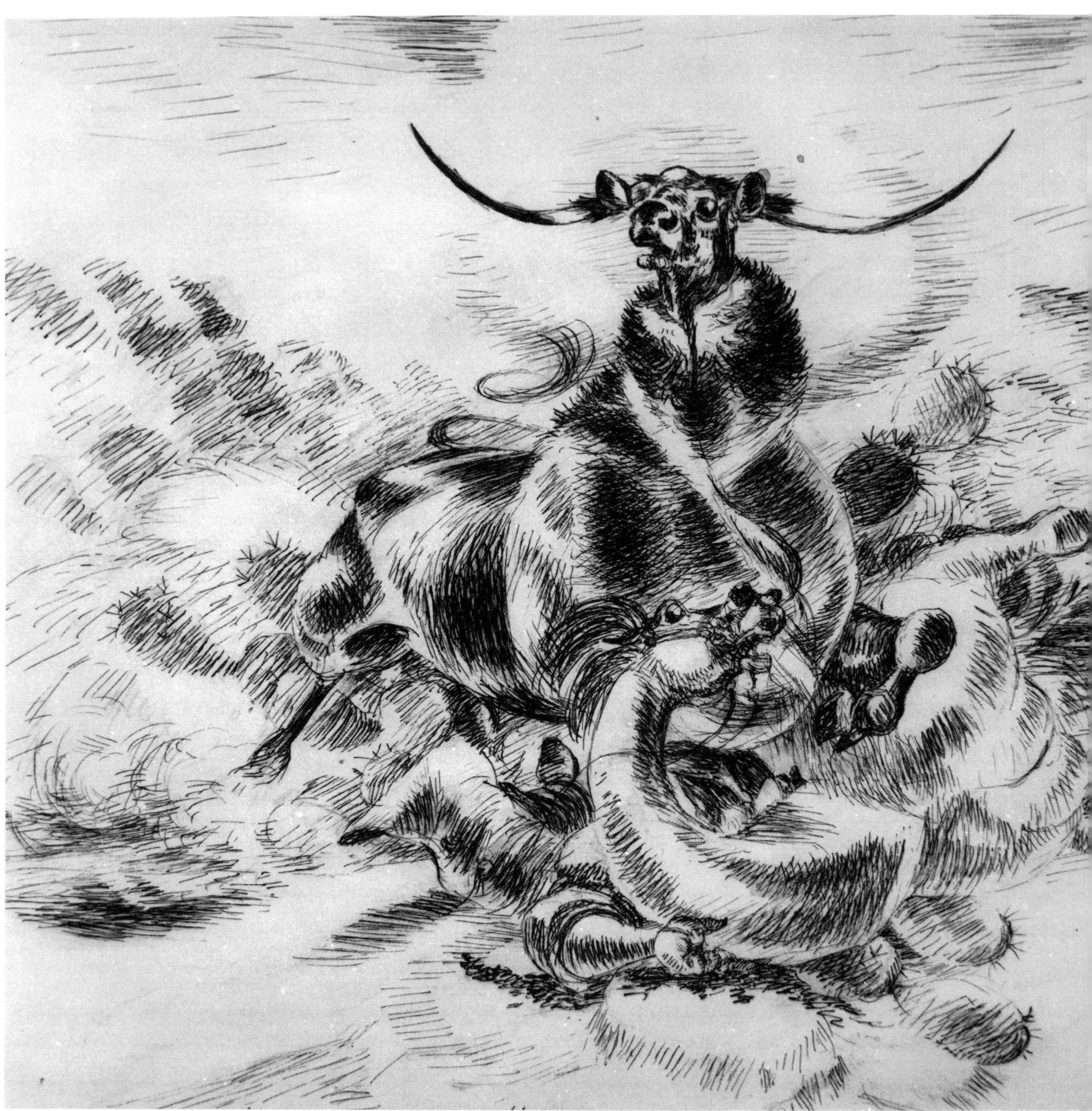

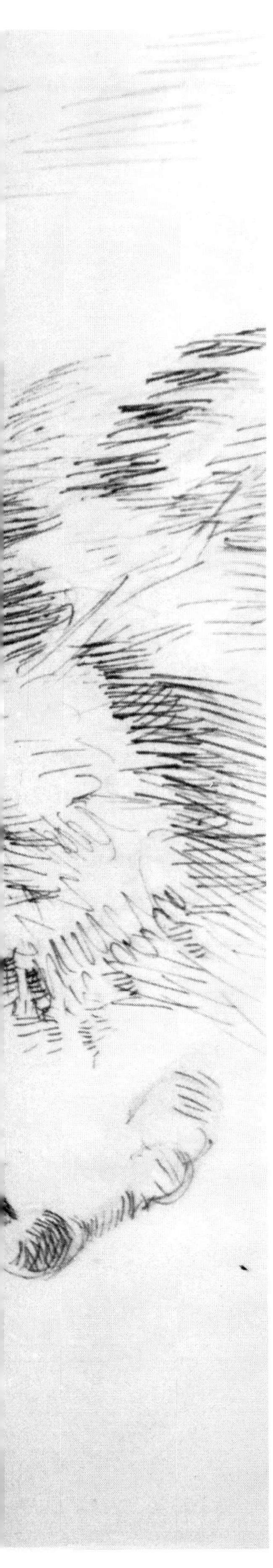

Revenge

Revenge! That was all a Mexico steer sought in Texas soil to live for,
To honor his pride with a duel to death with those
That at the end of a rope had brought him down.
A captive for the first time in this strange land
Where he had become the star of the wild,
Always the sentinel,
Always to lead the way of the free.

Never! Was he under the rule of man.
The only law he lived by was his own.
Free!
Even when he remained alone among his rivals, the Vaqueros.
To him, life, was not worth living in defeat.   —R.M. Beasley

DRAWING 21
El Mexicano. © R.M. Beasley Legacy LLC.
Reprinted with permission.

El Mexicano

HABÍA UN NOVILLO que nunca corría con otro ganado. Esto fué como en, el novecientos, más o menos, ocho o diez, por hay. En El Salerito, en el potrero El Carro.

Y este, este novillo era de Mexico al lado del río, al orilla de Mexico. A aquel novillo le decían "El Mexicano" (drawing 21).

Bueno, lo que pasó es que Sam Smithwick y otros vaqueros andaban lazando ladinos en "El Salerito" aqu/í al norte de San Diego. Y una mañana salieron, dice Sam y salieron muy de mañana. Y el sol amanecío allá en el potrero, y a poco andar, ellos sabían más o menos donde comían los ladinos. Ya no había quedado ganado, nomás el puro ladino.

Tenían que lazar.

Y llegan a un lugar en la orilla de un loma cerca de un ramadero donde habían estado hechados ocho o diés novillos. Y se acababan de levantar. Iban tumbando el sereno rumbo al ramadero. Y ellos los fueron sigiendo.

Siguiendo.

Siguiendo.

Y según, iban entrando al monte. Iban entrando más novillos. Pués fijate que cuando salieron a la otra loma, se pasaron al ramadero, ya iban diés y siete por la huella. Y se arrimaron a la orilla del monte, y los devisaron que andaban comiendo algo.

Este novillo era de Mexico, y era el más ladino de todos. Nunca lo habían podido pescar a este novillo. Desde que estabe aquí en Texas, no le había caido reata. Y dice Sam que, cuando vieron a los novillos todos, apiaron de los caballos y apretaron los silletes.

Iban a lazar.

THERE WAS ONE STEER that never ran with other cattle. Now that I think about it, it was around 1900 more or less, '08 or '10, around there. At El Salerito, at El Carro horse pasture.

It came from Mexico, on the Mexican bank of the river. They called it "El Mexicano" (drawing 21).

Sam Smithwick and others were roping novillos at "El Salerito" here north of San Diego. And they started out one morning, Sam says, and they left very early morning. The sun rose over that horse pasture, and a short ride later, they could tell more or less where the novillos grazed. There were no cattle left, purely wild novillos and nothing else.

They knew they had some roping to do.

Then they came to a place along the foot of a hill close to a brush pen where they had corralled eight or ten steers. They had just started. They were cutting through the fog toward the brush pen. And tracking all along.

Tracking.

Tracking.

And as they came into the brush, more steers kept coming in. In fact, when they came out to the other hillside, they passed along to the brush pen, there were already seventeen, judging by the tracks. So they approached the edge of the brush, and they could see them grazing something.

Now, this one, this steer was from Mexico, he was the wildest of all. They could never catch this steer. In all the time he was here in Texas, no lasso had ever fallen on it. Sam says that when they saw all those steers, they all dismounted and tightened their saddle chinches.

They were going to rope.

Y les dice el caporal, Ambrosio Montes, a Sam, dijó, "Tu te vas allá por lado del Mexicano," porque a aquel novillo le decían. "El Mexicano." "Al lado del río, al orilla de Mexico."

El novillo este estaba tan listo que no comía por estar viendo. Y a un tiempo se fueron mudiendo. La cuestión es que los quisieron encerar por todos lados pa' arrearlos para ciertos lugares que se fueran ya parados. Pero cuando vió el novillo a Sam que se hizo al lado de él, se le hechó en cima. Y agarró el monte. Y lo corrió como media milla. Una milla para cuando lo pudo lazar. Pero estaba tan liviano que así como lo tumbaba, se volvía a parrar. Y volvió hasta que se cansó el caballo. Así lo pudo amarrar. Y se fué al rumbo donde los campañeros habían ido con los novillos. Y los habían reparado. Estaban parados a distancia no muy lejos. Y le Dijó, le Dijó Sam, le preguntó el caporal, Ambrosio Montes a Sam, "¿Como te fué con el novillo, Sam?" Dijó, "Allá lo tengo amarrado."

"Bueno," Dijó, "Vamos mudiendo este de aquí pa' allá, y lo arrimamos cerca de donde esta pa' hecharlo a la trampa, para allá pa' el campo." Cuando llegan al lugar donde esta el novillo mexicano amarrado, van a el atajo de novillos que llevaban, y le dice Ambrosio a Sam, "Anda suéltalo, pa' traerlo paca."

"No," Dijó, "yo no lo voy a soltar." Dijó, "Ese novillo es muy bravo. Si lo soltamos, va matar un caballo, y hay vas." Dijó, "Anda sueltálo tú. Yo no lo suelto."

Entonces, mando Ambrosio a otros hombres a soltarlo.

Dice Sam, "Yo me retiré como cincuenta yardas y me escondí detras de un chaparro. Cuando lo soltaron, se levanto el novillo, y yo creía que iba atacar luego luegito. No. No atacó. Se puso a hechar tierra. Escarbar. Escarbar, y a hechar tierra. Garramando.

Y estubo un rato así, y de repiente, se puso grifo y se levantó y empezó a buscar alrededor algo que él quería hasta que devisó el caballo zaino, él que lo había lazado, él quería matar al que lo había lazado. ¿Me entiendes?" (Drawing 22.)

Y traía un caballo muy bueno, la marca veinte-cuatro Sam de aquí de Amargoso, los Pérez, esta gente Pérez del día. Este.

So the caporal, Ambrosio Montes, told them, he said to Sam, "You go on that side over there where the Mexicano is." That's what they called it. "El Mexicano." "Along the side of the river, along the Mexican border."

This steer was so alert that he wouldn't even graze, just to be looking around. So, for a while, they kept moving up. The point was that they wanted to round them up and drive them in from all sides to a place where they would be corralled. But when the steer saw Sam coming up to his side, it attacked. And ran into the brush. He chased it for over a half mile. A mile before he could rope it. But it was so agile that as soon as it was thrown, it would stand up again. So, he kept on roping until his horse was exhausted. That's the way he roped it. So, he went in the direction where his riders had herded the steers. They had them rounded up. They were herded at a distance not too far away. And he said, he said to Sam, the caporal, Ambrosio Montes, asked Sam, "How'd you do with the steer, Sam?" He said, "I got it tied over there."

"Good," he said, "Let's move them from here to there, and we'll close them in to put them in the pen, over there at the camp." When they arrived with the steers, where this one steer was tied, it was in the brush pen.

When they arrive at the place where the Mexican steer was tied, they go to the herd of steers they were herding, and Ambrosio says to Sam, "Go and untie it, to bring it over here."

"No," he said, "I'm not going to untie it." He said, "That steer is very fierce. If we let it go, it can kill a horse, and then you're in for it." He said, "You untie it. I won't untie it."

So then, Ambrosio ordered some other men to untie it.

Sam says, "I backed off about fifty yards and hid behind a brush thicket. When they released it, the steer reared up, and I thought it was going to attack right away. No. It didn't attack. It started kicking dirt. Digging. Digging and kicking up dirt. Pawing at the ground.

It kept doing that for a while, and suddenly, it flared up and started probing and looking around for something that it wanted until it spotted my chestnut horse—the one that had roped it. Understand? It wanted revenge" (drawing 22).

"Cuando lo ví que se dejó venir," [dice] Sam. "Saqué la reata," Dijó. "Que pase cerquitas. No. Ya que lo ví que venía derecho, le pegué al caballo y corrí estaba una loma." Y se le pegó el novillo. Y dice, "Yo sabía que entrando al monte, me alcanzaba. Y se me acabó la loma y entré al monte, y me alcanzó." Y se le emparejó el novillo a tirar una cornada en el corazón al caballo al codillo. Y Sam le dió una patada en los cuacos y le metío el cuaco por de bajo y lo hechó pa' afuera de la silla, y se fué colgado en el pesqueso. Y la otra estrope-ada, se la metió en el corazón y le hecho el caballo arriba.

Y mató al caballo.

Se voltea el novillo y viene a matar a Sam, y el caballo quedo con las patas pa' arriba que la sangre del caballo le estaba caiendo a Sam en el pecho y en la cara. Es cuando el novillo viene a cornar (drawing 22).

Pero cuando la res ve sangre se asusta. ¿Me entiendes? Se asustan al ver la sangre.

Vió la sangre. Empezó a garramar, a llorar como lloran ellos. Y se hacía pa' atrás. Venía atacar otra vez y miraba más sangre.

Sam no se movía. No se hacía como muerto.

Hasta que se le hizó por detrás y la sangre que tenía aquí en la frente, con el pico se la tallaba y le lloraba garramando en la frente.

Y Sam no se mueve.

Entonces, de repente sienten el novillos los compaños, Sammy Doss, el papa de Pedro, se había venido detrás de Sam, siguiendolo. Y cuando lo vió al novillo. No. No lo sintió al novillo luego.

Vinó y vió que el novillo estaba arriba de Sam parado arriba de las manos y en la cara de él. Y le veía la sangre. Fué y aviso.

Dijó, "A Sam le mató el caballo y lo mató a él también. Está lleno de sangre, y está el novillo arriba de él. Sangrien-tos las manos así."

Entonces vinieron los vaqueros. Cuando lo vieron al novillo, tenía la lengua de fuera, hechando babas. Y cuando el novillo los sintió, fué y los topo. Fué y los encontró.

He waited there for 'em (drawing 23).

Y los hechó en corrida, y cornó otro caballo.

And that was a very good horse, the "Twenty-Four" brand that Sam was riding, from over here at Amargoso, the Pérez family, these present-day Pérez folks north of Alice.

"Well, when I saw that he cut loose toward me," says Sam, "I took out my lasso," he said. "Let him come close. No. Actually, once I saw him coming head-on, I spurred my horse to run, toward a hill." And the steer was close on his heels. So he says, "I knew he'd catch up with me in the brush thicket. But it was a small hill, so I quickly found myself in the brush, and he caught me." The steer lowered its horns, going for the horse's heart behind the shoulder. So, Sam kicked at the horns, but it drove its horns up and threw him out of the the saddle, and he went over, clinging to the neck. And the next thrust, it gored into the heart and threw the horse over his shoulder.

So, it killed the horse.

Then, the steer turns on his heels and comes to kill Sam. Well, the horse was lying, hooves skyward, with the horse's blood flowing on Sam's face and chest. That's when the wild steer comes in for the kill (drawing 22).

But when a steer sees blood, it's frightened by the sight of blood. It's hard to understand.

Well, it saw the blood and started pawing and bellowing the way they bellow, and it drew backwards. It was coming to attack again and saw more blood.

Sam wouldn't move. No. He played dead.

And then it started backing up. And the blood it had on its forehead, it kept trying to rub it off. It kept smearing it with his snout and bellowing and pawing its forehead.

And Sam doesn't move.

So, the other vaqueros suddenly noticed the steer. Sammy Doss, the father of Pedro, had been coming along behind Sam, following him. And when he saw the steer . . . Actually, he didn't notice the steer right away.

But when he came closer, he saw that the steer was standing over Sam—standing over his hands and his face. And he saw all the blood. He ran off and reported it.

He said, "He killed Sam's horse, and it killed him too. He's covered with blood, and the steer is on top of him. Blood all over him."

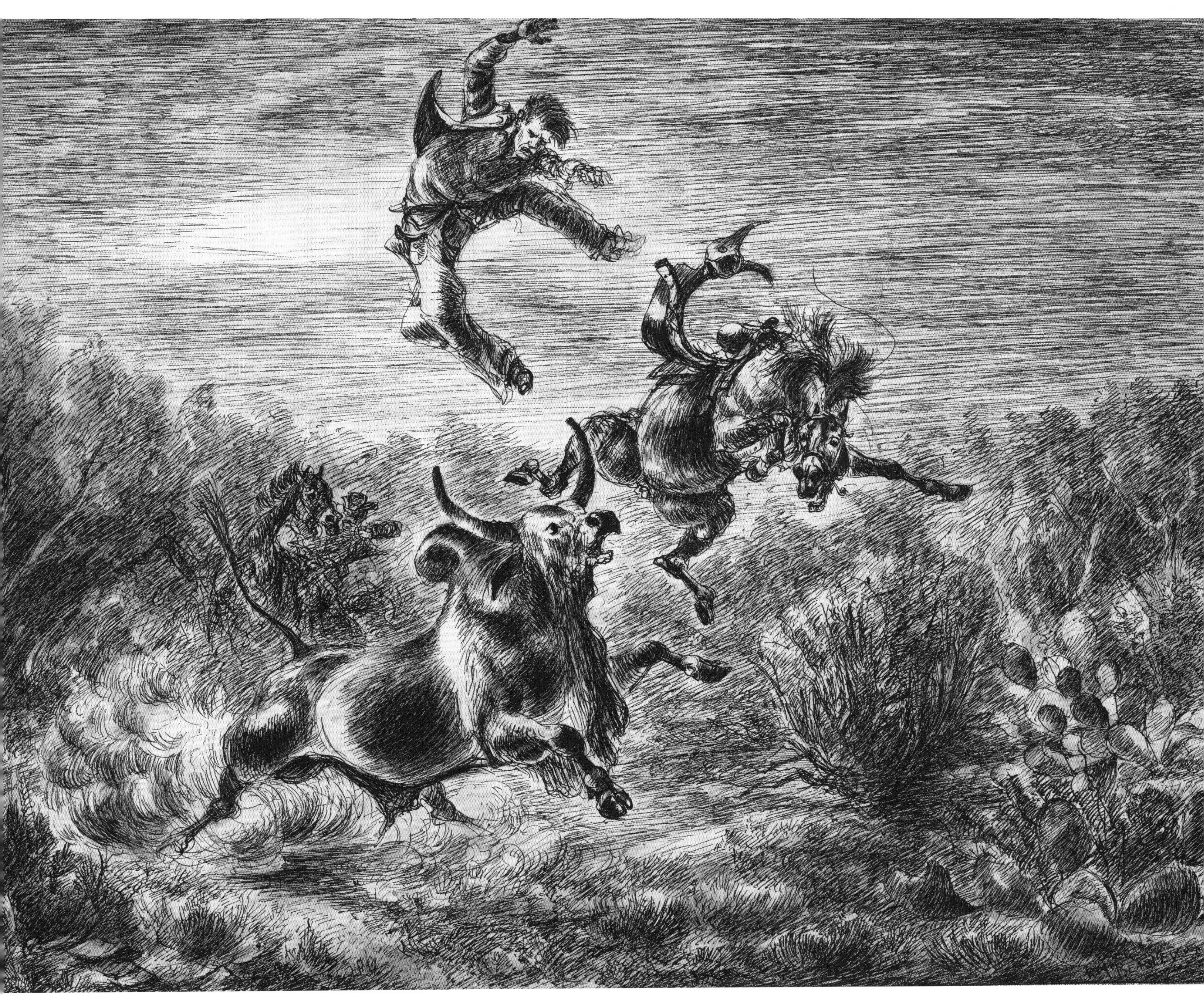

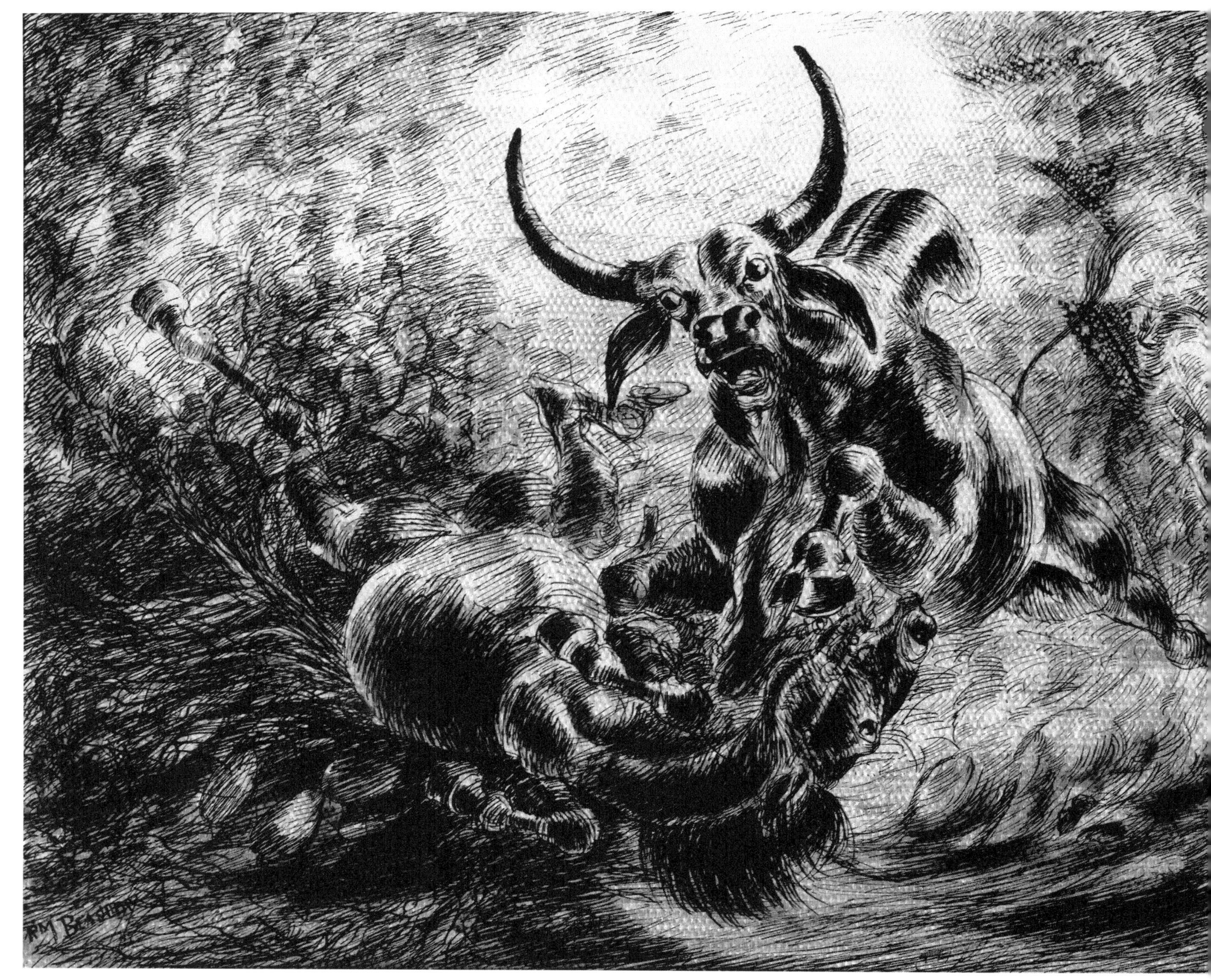

Y ya no volvieron. Ya no hubo quien volviera ya viendo aquel animal. ¿Quién se arrimara?

Pero en ese instante en que fué el novillo corriendo los vaqueros, Sam se pudo salir porque había caido entre ramas el caballo. Se pudo salir, y fué y se subió al mesquite. Sam lo vió con el cuaco lleno de sangre, dijó, "Ya mató otro él." Se quedó arriba del palo y no se movió.

El novillo lo buscó porque la res busca pa' arriba. Lo vió, y como Sam no se mudía, allí estubo un rato más viendo el caballo, viendolo al él. Al fin, dice "Oyes," y se iba llendo poco a poquito y volteaba pa' atrás, a ver quien quería más pleito. "¿Eh?" Y así se fué poco a poquito hasta que se perdió en el monte. Y Sam nunca les gritaba a los compañeros porque si gritaba, una res enojada, gritandole vuelven. They come back. Lo dejó pasar por una hora. Que se fuera lejos. Y entonces sí gritó.

Al grito, los vaqueros no podían comprender quien estaba gritando. Porque ellos lo creían muerto. ¿Entiendes? Ellos no esperaban de que fuera Sam.

"¿Pues quién será él? ¿Quién está gritando?" Y se vinieron a ver que. Ya cuando lo hallaron, arriba del palo.

Se quedaron almirados. Y lo que es el espíritu de un vaquero de esta clase tan valiente, tanto así como el toro. El novillo tenía la venganza de matar al caballo y a Sam, a Sam le quedó la venganza de que el novillo lo había ganado después de todo.

Le quitó un caballo prestado a uno de los vaqueros. Fué y lo alcanzó y lo lazó otra vez. Y cuando llegan los vaqueros, le dice Sam al caporal, "Vamos a amarrarlo a un palo, y traernos un buey, y que el buey lo lleve."

"No," le dijó, "Lo ponemos entre los reatos de la pata y el pescueso, y lo llevamos."

"No," le dijó Sam. "Revienta una reata y mata los caballos."

Pero no le hizó caso. Pués le pusieron como quería el caporal. Reventó una reata y fué y cornó otro caballo. Y ya no sé que quien estubo más. El animal. Allí se acabó.

Tuvo, la cuestión es que el novillo . . . Mató tres en un día, tres caballos (drawing 24).

So, the other vaqueros came. They saw the steer with its tongue hanging out, frothing at the mouth. And when the steer heard them, it went and butted them. It went up to meet them.

He waited there for 'em (drawing 23).

Well, as it ran them off, it gored another horse.

So they didn't return. Nobody would go back, after having seen that animal. Who'd go near it?

But at that moment while the steer was chasing the vaqueros, Sam was able to escape because his horse had fallen in the brush. He got out and climbed a mesquite. Sam saw it and noticed the blood on its horns and said, "Now he's killed another horse." He stayed up in the tree and didn't move.

The steer looked for him because cows look upward. It saw him, but Sam didn't move. So, it stalked around a while longer, looking at the horse, looking at him. Finally, as if to say "Hey," it started backing off little by little, and looking back to who else wanted a fight. "Eh?" That's the way it kept backing off little by little until it disappeared into the brush. But Sam never shouted out to his companions because if he did, an angry steer would come back to the shouts. They come back. He waited an hour after it left. To let it go far away. Then he did shout out.

But at his shouts, the vaqueros couldn't understand who was shouting. Of course, they thought he was dead. They didn't think it could be Sam.

"Well, who could that be shouting?" they wondered. So, they came to see. That's when they found him.

They were amazed. And what is so valiant about a vaquero with so much spirit, as much as the bull, the steer had a vengeance to kill the horse along with Sam, and Sam held a vengeance against the steer that had, in the end, defeated him.

So, Sam takes a horse from one of the vaqueros. He went out and caught up with El Mexicano and roped it again. So, when the other vaqueros came back, Sam says to the caporal, "Let's tie him to a tree. Bring up an ox, necktie them, and let the ox lead him."

Mató un lio. Pero era un novillo que nunca desde que vino de Mexico, nunca lo habían podido pescar.

Mira, pregunto yo a Sam como estaba platicando—porque yo todos mis dibujos los hago basados en los movimientos que hace el vaquero cuando está platicando—de "¿Que clase de llaves tenía?"

Dice así. Por eso lo pudo pegar en el corazón. "Si haya estado pa' arriba no lo cuerna. ¿Eh?" Tenía que tenerlo extendido.

"No," the caporal said, "We'll hobble him foot and neck, and take him off."

"No," Sam told him. "He'll break the rope and kill the horses."

But they didn't listen. So, they hobbled it like the caporal said to. Well, it broke the rope and went and gored a third horse. And I don't know who got the most out of that day. The beast? It ended there. It ended.

The point is that the steer . . . he killed three in one day, three horses (drawing 24).

He killed a bunch. But it was a steer that, not since it came in from Mexico, were they ever able to catch it.

I got to ask Sam as he was talking—because I always based all of my drawings on the motions that the vaquero makes when he's talking—asked, "What kind of horns did it have?"

And he says, "Like this . . . ," motioning both outstretched arms forward. That's why it could gore a horse in the heart.

"If they had been pointing outward, it couldn't have gored it so deeply. Huh?"

It's because they extended forward.

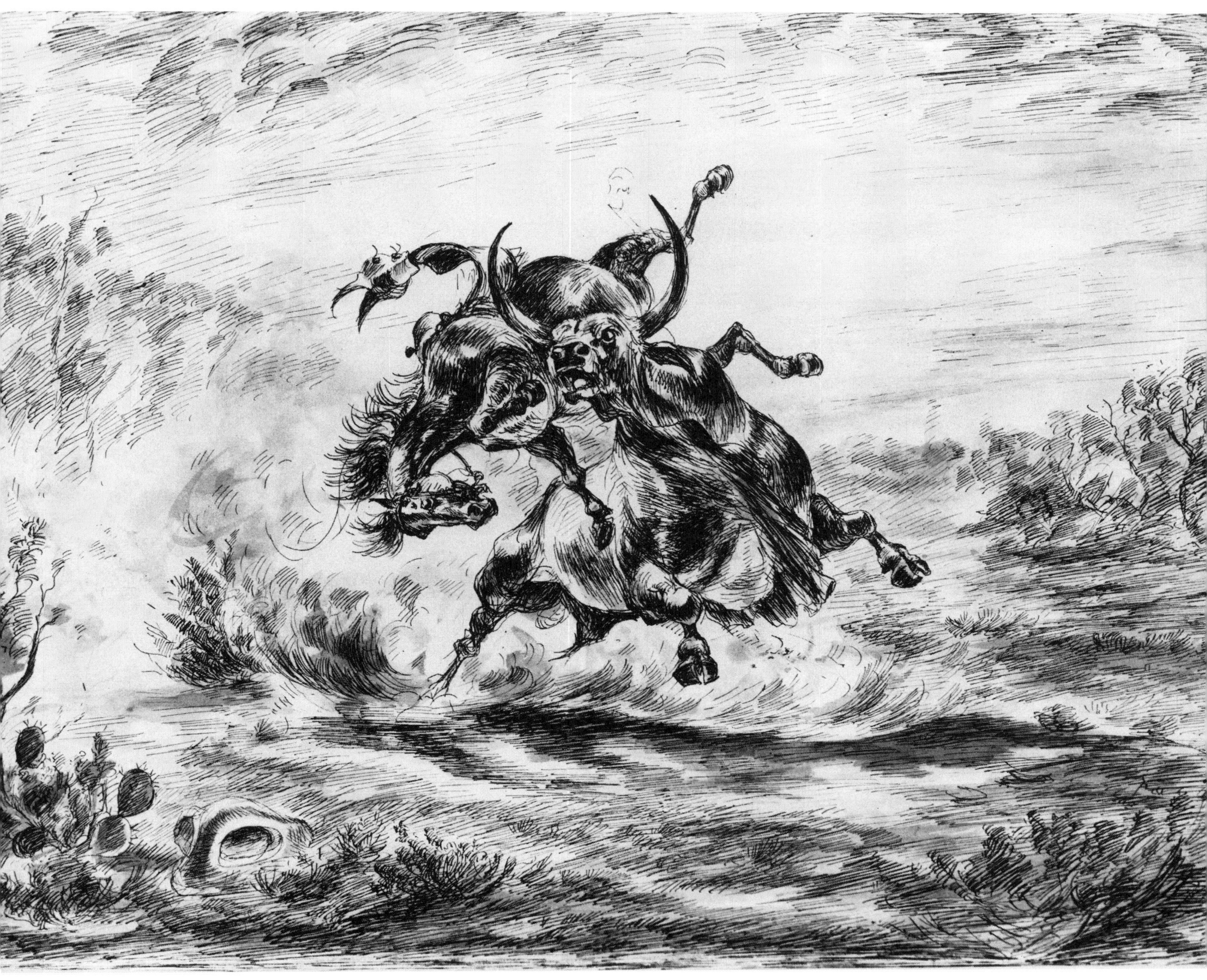

DRAWING 24
It gored a third horse. © R.M. Beasley
Legacy LLC. Reprinted with permission.

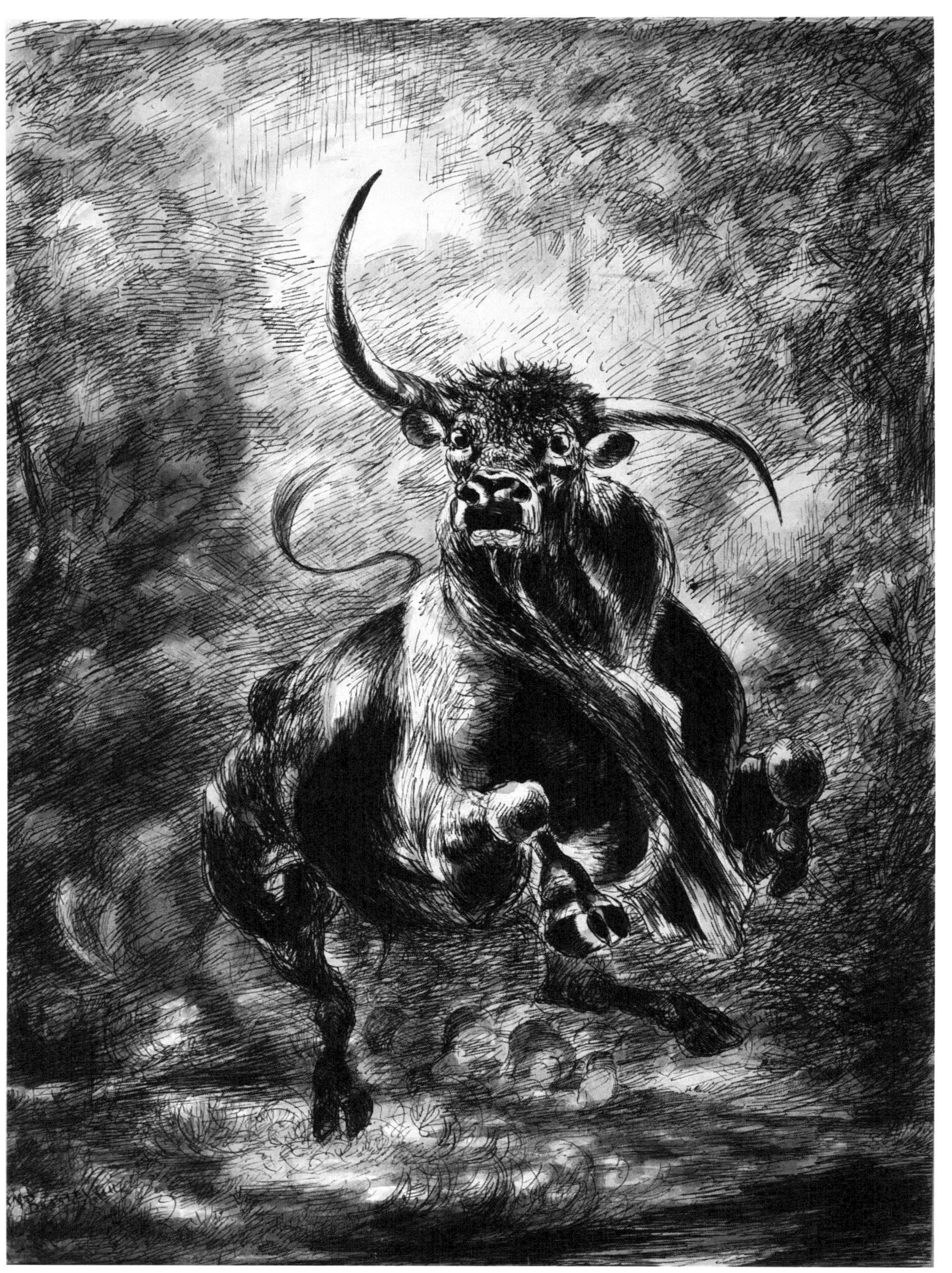

DRAWING 25
El Gacho. © R.M. Beasley Legacy
LLC. Reprinted with permission.

The Magnificent Rebel

Mad with hatred, was Gacho,
The Savage killer bull of the Nueces River,
Remnant of the Longhorns,
Untouchable, never possessed,
His life had been a duel to death
With those who sought to crush his spirit . . .
His right to be free,
Never once turned his back to the enemy.

"Better dead than a captive"
His cry of war seemed to be—
Than to exist only and no longer live
But for the worst of endings in humiliation.
The longest of agonies, the cruelist of means,
To drain life and remain alive . . .
More infamous than by a drop of blood,
Slowly, one drop at a time.

Thus, lamented The Magnificent Rebel,
The anguish of such a doomed existence . . .
He who had known the glory of freedom
Ever since the light of the universe
Alighted his birth and enlightened his life
The rest of the way.
Ricardo Moreno Beasley, 1974

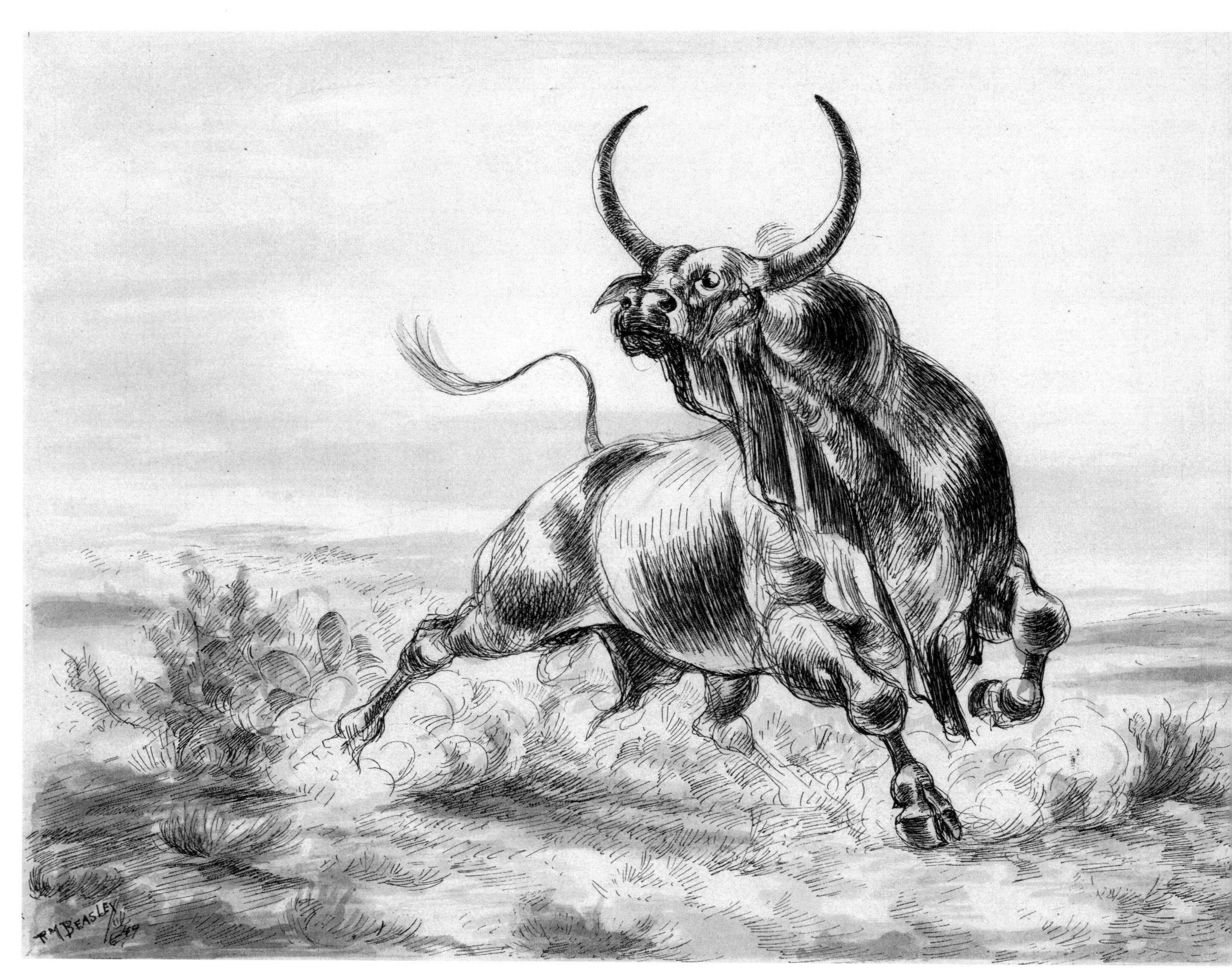

El Zermeño de Duval

EL ZERMEÑO. This was a monstrous steer.

El Zermeño era de Duval. Aquí nació en Duval en Siete Hermanas. Era mancito. Comía pan allí alrededor de las casas allí cuando tenía como un año o dos. Cabrecía como un caballo.

Y luego lo vendieron los Sernas a Don Manuel Rogers, cuando tenía tres años se lo llevaron para La Campana. La Campana está como cuarenta millas al norte de aquí. Allá cuando lo hecharon allá, sería por el sentimiento de haber perdido su casa aca. Los animales, han de tener algun amor de hogar porque nunca quizó juntarse con nadie. Vivió solo todo el tiempo. Y fué creciendo tan grande que se hizo muy peligroso que no salía de día, nomás de noche.

Decía Teófilo, como dice por la manera como lo ponía él, "Era como un bandido El Zermeño." dice, "Nomás de noche salía." Y andaba muy poquito. Y la cuestión que en toda la vida que estubo allá, lo hecharon pa' allá en novecientos cinco, murió, lo mataron en veinte y tres, duró allá, este . . . Todos esos años en ese lugar, pero lo raro que fué, hizo su casa porque los animales, ellos también hacen su casa. Y tienen sus modos que ellos allí viven diario.

Una vez, Teófilo buscando, siguiendo jabalinas, un día fuí yo con él en la caza de novillos. En un rincón que le decían La Piedra de Lumbre. Tenía allí el novillo los palos rosados de muchos años de estar allí, pero fué y hizó su casa lo más cerquita que se pudo arrimar rumbo así para el mar, a Duval. A esa esquina se escarbó.

Y era tan—Teófilo le siguió la huella—como caminaba. Salía por un lugar a comer y entraba por otro. Nunca entraba por el mismo, pa' venir, y andaba la manera anda él. Andaba un poquito pa' alla, y luego pa' tras sobre la huella.

EL ZERMEÑO. This was a monstrous steer.

El Zermeño was from Duval. It was born here at Seven Sisters in Duval. Well, it was very tame. When it was one or two years old, it'd eat bread around the houses. It nodded its head like a horse.

And then the Sernas sold it to Don Manuel Rogers when it was about three years old, and they took it to La Campana. La Campana is about forty miles north of here. But over there, when they put it out over there, maybe it was because of its hurt feelings over losing its home over here. Cattle must have some love for home, because it never again sought company. It lived alone, always. So, it kept growing so large that it became very dangerous and it wouldn't come out by day, just by night.

Teófilo used to say, the way he put it was, "He was like a bandit—El Zermeño." He said, "Only by night would he come out." And it ranged out very little. Well, the point is that during its whole life that it spent there, they put it out there in 1905, he died, they killed it in twenty-three. It stayed there, let's see . . . all those years in that one place, but what's so unusual, he made his home there because animals also make a home. And they have their own way of living, and they live there daily.

One day, Teófilo went out, hunting javalinas, I went out with him to hunt wild steers. In a remote meadow that they called the Rock of Fire, this steer had all the trees rubbed raw after being there for many years, but he had gone and made his home as close as he could get in the direction of his old home, in Duval. Hard to believe. In that niche, he nestled in. And he was so—Teófilo followed his tracks—trailing it. It'd go out from one place to graze, but it'd enter at another place.

Y luego pa' aca y luego pa' el otro lado. Así cuando lo iban siguiendo, nunca lo van a hoder porque todo el tiempo te sentía. Estaba más allá.

O estaba pa' tras. Smart animal. Muy inteligente. Muy inteligente.

Teófilo lo refería el animal más inteligente que una gente para cuidarse. Déjame acordarte de la inteligencia de él.

Dice Teófilo que él un domingo fué a las jabalinas, a los venados, y estaba llovisnando, y estaba muy nublado el cielo, y para no perder el caballo, lo dejó en la presa de Montalvo, un lugar muy conocido. Allí lo dejó. Y se fué rumbo al norte. Iba siguiendo las jabalinas, la huella, cuando oyó tronar pencas de nopal.

Y dijó, "Seguro que son las jabalinas."

Y se fué a gates a gates hasta que vió que eran tres novillos que estaban entre un nopal. Y en el medio estaba este. Al tiempo que él lo mira—él salía tanto, estaba arriba de los otros—al tiempo que él lo mira, se le vió la marca "HR" en la pierna. El novillo trozó una penca y levantó y así pesco.

En un rato, pesco el aire y bufó y se fué. Corrió rumbo al norte porque ya tenían la orden que lo mataran. Era muy peligrosisimo. Ya había matado un caballo, herido dos. Era muy peligroso.

Entonces dice Teófilo, "Ahora sí lo mato. Va rumbo al norte." Y vió que estaba llovisnando. Lo iba siguiendo. "Pero pues muy bién," dijó, "Ahorita me lo hallo." Cuando de repente, le bufa aca atrás.

Ja ja.

De el otro lado.

Fué y le cortó atrás para que mire. ¿Me entiendes?

Y dice, "Así me trajo toda la tarde. Bufandome a una distancia nomás que no lo viera," pero nunca lo pudo alcanzar, ni ver ni a lazarlo.

El cazador se volvió el que andaban cazando—el que andaban cazando. Lo cortó tres veces.

Pero dice Teófilo, lo burló, "Que andaba haciendo burla de mí toda la tarde anduvo bufando, y nunca lo podía encontrar."

Daban una descripción que daban ellos de los cuacos de ese.

It never entered the same place, to come back in. And it walked—the way it walked was— it'd walk a ways over there, and then it'd backtrack over its own trail. And then back this way, and then back the other way. So, when they tracked it, they could never fool it because it could always sense you. It'd be farther down.

Or it'd be behind you. Smart animal.

Very intelligent. Very intelligent animal.

Teófilo used to say that this beast was smarter in self defense than any human. Let me give you an idea of his intelligence.

Teófilo says that one day, on a Sunday, he went after javalinas, after deer. It was misty, and very foggy skies, and so as not to lose his horse, he left it at Montalvo's dam, a well-known place. That's where he left it. So, he set out to the north. He was tracking the javalinas, their tracks, when heard a crunching of cactus pads.

He said, "It must be the javalinas."

So, he goes crawling, crawling on his hands and knees until he could see that it was three wild steers in a cactus patch. And that one steer was right in the middle. And just as he saw it—it stood so tall, it stood well over the others— just as he saw it, he could see the "HR" brand on its leg. The steer ripped a cactus pad and lifted like this and caught it.

In a split second, he caught a whiff, and let a snort, and took off. He ran off to the north. Now, there was a standing order to kill him. He was very extremely dangerous. He'd already killed one horse, wounded two. He was very dangerous.

So then, Teófilo says, "Now I'm going to kill him. He's headed north." Well, he saw that it started misting. He was tracking it. "But well enough," he said, "I'll find it anyway." When suddenly, it snorts at him from behind.

Ha ha.

From behind him. It backtracked and came up behind him. Unbelievable.

He says, "It kept me going like that all afternoon. Snorting at me from a distance just out of sight." But he never could catch up with it. Neither to see it, nor to rope it.

The hunter became the hunted—the one that was being hunted. It backtracked three times behind him.

Decían ellos de él que tenía unas llaves encornadura de toro de pelea. De allá le tiran para abajo de la base, y luego le levanta como canasta. Una llave muy peligrosisima.

Bueno, estos novillos son parte de longhorn. No tienen cuaco del longhorn, como el longhorn. Porque el longhorn había todo otros shapes. Pero pertenece a ese breed.

And the size, . . . muy grande. Muy grande.

Este, ese cuadro le quiero arreglar, ponerle el mat nuevo a ese dibujo. Que bonito.

Este es uno que enseña un toro de una proporción muy grande. Pero no se ve esa más grande. Esta comiendo en la punta del nopal. No se ve ese más grande que si estuviera parado, que se viera desde abajo.

Y esos novillos que andavan así, podían existir sin agua. Sin agua. Nomás comiendo el nopal.

Puro nopal.

"Y eran tan vivos," dice Peto Mendoza, "pa' ir a comer el nopal, usaban el tronco porque tenía más juice. " Pero antes que llegar al nopal que tenía un tronco grande, le entraba por este lado y se reculaba y entraba por el otro y le entraba. Cuando estaban todos los lados seguros que no había peligro, entonces se arrimaba. Le daba un cuacazo al tronco y sacaba un taco. ¿Me entiendes? Un taco de nopal, y lo mascaba, y a puros cornados.

Y aun eso con que podían existir con nopal, este fué caiendo de una seca. Lo pescaron al Zermeño cuando hubó seca. Se le acabó el agua . . .

Ni el nopal tenía agua. Se vino colando por un arroyo buscando agua, y llegó a La Campana. Y a Peto lo tenían cuidando una puerta de la presa que caieran ganados del dueño del rancho. A este no lo esperaban. Cuando calló este animalón.

Decía Teófilo que una vez, llegaron a la orilla del monte, y estaba como cuarenta yardas el corral. De allí no a lo precauvido para huirse, de allí de la orilla del monte a la puerta, llegó como a las cuatro de la tarde en verano.

Para llegar de allí le obscureció y no llegó a la puerta.

No se llegó.

Cuando se iba a media pesuña, y cuando oyía un ruido—brincaba una rana en el agua or algo—se reculaba para atrás. Nunca daba la espalda para correr. De recula así,

And Teófilo says it teased him, "He was just making fun of me all afternoon, just snorting, and I never saw him."

They used to describe that steer's rack of horns.

They would say about him that he had the most powerful rack of long horns. The horns dropped way down from the base at his skull, and then they curved up like a basket. A very dangerous rack of long horns.

Well, these steers were part longhorn. They didn't have the longhorn's rack, like a longhorn. Because a longhorn had all types of shape. But it belongs to that breed. And the size, . . . very large. Very large.

Well, I want to frame that drawing, use a new mat and put a mat on it. How pretty.

This one shows a bull of large proportions, but it doesn't really look bigger. It's nibbling on the tip of a cactus plant. It doesn't look as big as if it were shown standing erect, as if you were it were shown from below.

And those steers that were like that, they could go without water. With no water. Just by eating cactus. Just cactus.

And they were so clever, says Peto Mendoza, "When they ate cactus, they would eat the trunk because it had more juice. But even before getting at a cactus plant that had a large trunk, they would approach it from one side, back out, and approach it from another side before going in. Only after seeing that there was no danger on any side of the cactus, did they go in. They'd gouge out a big taco of cactus with one swing of their horns at the trunk. Hard to believe. A cactus taco, and they'd chew on it, just by gouging it out with their horns.

And with all that being able to survive on cactus, that was because of the coming drought. They caught El Zermeño because of a drought. The water dried up . . .

Even the cactus dried up. It came passing along a gully in search of water and arrived at La Campana. And Peto was assigned to guard the gate to the pond of the dam, watching for the ranch owner's herd to come in. They didn't expect this one. When all of a sudden here comes this great beast.

Teófilo used to say that once it came up to the edge of a thicket, and the corral about forty yards away. But it approached cautiously, ever on the alert to flee, coming up from the edge of the thicket to the gate. It arrived at about

viendo todo el tiempo el peligro. Y era tan observante que en todo se abajaba él. Si un pajaro volaba, se fijaba por qué voló el pajaro. Si un ratón corría, se fijaba a ver qué lo asustó al ratón. Ponía observaciones como una gente. Más que una gente.

Por eso pudo veinte años de existir así sin que lo pescaran.

Ahora, se trataron de matar con carabina y nunca lo alcanzaron a matarlo. Y es mucho sentido, muy inteligente.

Y este fué uno que también alcanzó matar un caballo, aunque a vaquero no mató.

Mató, mató El Cubanito, un caballo.

El novillo, lo lazó un Americano de allá del . . . de La Campana. Y lo aventó con todo y reata.

En la, fué en la mañana, y en la tarde fueron a buscarlo.

Iba entre ellos un tal Donato Colón. Dijó, "Yo lo sigo," cuando entró al monte, "Y ustedes se van y se hacen allá a la punta del llano." Y lo iba siguiendo cuando, por la rastra, cuando se halló la reata. Iba mal lazado. Y a poco, levantó al novillo y les gritó a aquellos. Fué y le Salió cerquita pero le tuvieron miedo porque lo dejaron que llegara al otro monte, y ellos corriendo aca de lejos. El novillo llegó al monte y los esperó. Iba a la casa de él a donde fué a dar, era la casa de él. Y los esperó a la orilla del monte.

El primer caballo que llegó, lo mató en el corazón. Al otro lo pescó de arriba de el codo, le metió la llave hasta el tronco arriba del codo. Se le secó la pierna al caballo.

Y al vaquero donde lo llevaba arrempujando así con la cornada, un ramo lo tumbó y le caió para abajo, pero el novillo, como era enfurecido del caballo, no le puso cuidado. Si no, lo mata a él también.

Se le fué por arriba.

Pasó por arriba.

Damn. Pesa unos dos mil y ahora que te pasan por arriba sin . . . Ese animal se mira muy bonito de lejos por la sombra, de lejos.

Que lástima que lo tuvieron que matar con un tiro.

Que lástima. lástima.

Y dice Peto que fué tan valiente, esa tarde que lo vió el llegar, estaba espiando él, llegó el patron, el caporal de Robstown, Miller Morsby. El novillo cuando sintió el ruido,

four o'clock that summer afternoon. But it got dark just as it approached, so it never entered the gate.

It never entered.

When it walked at half-a-hoof, or when it heard a sound—whether a frog jumping in the water or anything—it quickly backed away. It never turned its back to run away. Backing away like that, always facing toward the danger. It was so observant that it stopped for anything. If a bird flew, he stopped to see why the bird flew. If a mouse ran, it stopped to see what frightened the mouse. It was as alert as any person. Even more than a person.

That's why it could elude capture for twenty years.

Now, they would try to shoot it with a rifle, but they never could catch it. It was very sensitive, very intelligent.

And this was the one that eventually killed a horse, but it never killed a vaquero.

It killed El Cubanito, a horse. This steer, it had been roped by an Anglo American from over at the . . . over at La Campana Ranch. Well, it pulled him off his horse, rope and all. That was in the morning. So that afternoon, they went out to look for him.

And riding among them was a certain Donato Colón. He said, "I'll go in and get him," so he went into the brush, "And you all go and form along at the end of the clearing." So, he went following it, by the trail of the rope that it was dragging, when he found the rope. The lasso had been poorly thrown. Shortly, he caught up with the steer, and he shouted out to the others and came up close to the steer. But he had to . . . because the others were afraid of it and let it escape into another thicket, and they were following it only from a distance. The steer got into the thicket and laid in wait for them. It had gone into its home thicket. It was his home. So, he was waiting for them at the edge of the thicket.

The first horse into the brush, it gored to death right in the heart. The next one, it caught above the elbow, driving its horns in to the hilt. Above the horse's elbow, severing the leg.

And as the horse was being shoved and gored with the horns, the vaquero was knocked over by a branch and fell underneath. But the steer was so infuriated at the horse that it didn't notice him. Otherwise, it would kill him too.

It just stepped over him.

arrancó. Y cuando llegó, le dice Peto al caporal, dijó, "Hay
está el Zermeño. Viene encalmado."

Dijó el caporal, "Traete la carabina." Y fué al rancho y
trajó la carabina, y ya la tarde caió. Y entonces así encalmado, el novillo se dejó ir derecho al agua. Y cuando entro
adentro, le tiró aquel, y le dió un tiro en el tronco a la llave.

Asonzado fué y le pegó a la cerca y se llevó la cerca y
se fué. Y por detrás, le dió un tiro en el pescueso. Y caió.
Un tiro muerto. Y dice Peto que caió incado y tres veces se
levantó y volvía incar y volvía a caer. Fué un tiro al alma,
hasta que caió tres veces.

Tres veces se levantó.

Mucha fuerza.

Cuando lo mataron lograron mucha carne.

Y tres años de manteca.

Pesaría unas qué, unas dos mil libras, verdad.

El novillo, sí. Era muy grande.

No sé qué más e puedo platicar de este novillo.

De este novillo, lo habilidoso, lo inteligente que era.

Decía Teófilo, "Para cuidarse cuando iba al agua, como
todo el tiempo tenía sospecha él de que hubiera peligro,
seguía a otro ganado. Entiendes?

Nunca solo.

No. Que fueran otros primero. Si los otros volvían, se
arrimaba él. ¿Entiendes?

Pues, una noche estaban espiando en un papalote arriba
de la mesa y fué un alambre estirado a la puerta. Llegaron
unos novillos y entraron al agua y este se quedó afuera. Ellos
esperaban que él se metía pero este no se metío. Se quedó
esperando a ver si aquellos salían. ¿Entiendes? Si aquellos
volvían, entraba él. Los novillos estaban tomando agua muy
recio, pronto.

Y le dice Teófilo a los compañeros, "Sierra la puerta
porque al cabo esos novillos se van a ir tambien," les dijó, "y
aquel no entra." Cuando le iban serrando a la puerta, estaba
parado en la puerta así.

Se paró el macho. Volvío y se fué.

Se tuvieron que tomar la decisión que si iban a pescar
cinco o seis o nomás uno.

Pero El Zermeño no entra. Le digo que tenía otra precaución para evitar peligro. No tomaba agua en arroyos.

Passed right over him.

Damn. It weighs over two thousand and now it steps right
over you without . . .That beast looks magnificent from a distance in the shade, from far.

What a shame they had to shoot it to death.

It's a shame.

And Peto says that it was so valiant, that afternoon when
he saw it come in. He was scouting, and the boss arrived, the
caporal from Robstown, Miller Morsby. And when the steer
heard the noise, it took off.

So, when he got there, Peto tells the caporal, he said,
"There's El Zermeño. It's coming in very dehydrated."

The caporal said, "Bring the rifle." So he went and brought
the rifle from the ranch. It became late afternoon. So, as
dehydrated as it was, it went straight for the water. When it
went into the water, he shot it. He shot it right at the base of
the horns.

Stunned, it went into the fence. It ripped the fence away
and ran off. So, shooting from behind, they put a bullet right
in the neck. And it fell. One lethal shot.

But Peto says that it fell to its knees, and three times it
stood erect and back on its knees. It kept falling. It was a bullet to the soul, that it fell three times.

Three times it stood up.

A lot of power.

When they shot it, they got a lot of meat. And three years
of lard.

It must have weighed two thousand pounds.

That steer. Yes, it was very large.

I don't know what else to you tell you about that steer.
About that steer, its great ability, how smart it was.

Teófilo used to say, "Its way of guarding itself when it went
to water, since it was always so fearful that danger may be
lurking, was to follow behind other steers." Hard to believe.

Never alone.

No. Let the others led the way first. If the others could go
in and come out, then he'd approach. That was it.

Well, one night they were scouting, sitting on the platform
at the top of a windmill. They had a wire stretched to the
gate. Two steers came up and entered to drink water. That
one steer stayed out. They thought he'd go in, but he wouldn't

Tomaba agua en charcos donde no hubiera gente que se arrimara. Entiendes? Charquitos aquí, charquitos allá. Evitaba tomar agua en un lugar donde fuera frequentado por la gente. Mira, y venía a tomar.

Lo raro es que hacerce tan ladino, tan salvaje, siendo un mancito aquí que comía tortillas, cabeciaba con mecate. Cosa rara. Y hacerce tan peligroso.

Sentería el novillo que lo habían denunciado.

Pues yo creo que resintió que lo habían quitado su hogar, y se aisló. Le quitaron sus compañeros. La gente de él, en otras palabras. ¿Me entiendes? Ye él se sintió muy serio. Digo yo que sentería eso porque nunca quiso compañía, ir con alguien. Lo mismo que para crusar un llano. Si había otro ganado que pasara adelante, él se venía atrás. Que pasaran otros. Y otra cosa, no andaba por vereda tampoco. El abría el monte solo.

Solo, él. Él no seguía vereda de nadie. Decía Teófilo, que es una cosa rara de este animal. La inteligencia. Es más inteligente, la razón que el novillo de seguro tanto la inteligencia es como no tenía sex. ¿Me entiendes? No había aquella necesidad de salir como la vaca o el toro que salen por la necesidad que los obliga la naturaleza. El novillo no. Usaba todo, no había tentación. Usaba toda su inteligencia en cuidarse. Y hubo otros como este que andaban así, perdidos, o que no los podían buscar ni hallar. Había otros así como este.

¿De novillos? Sí. Hubo varios.

go in. He stayed out, waiting to see if the others would come out. Amazing. The steers were drinking water very fast, quickly.

So, Teófilo calls out to the fellows, "Just close the gate anyway or those other steers will get away," he told them, "that one's not coming in." When they pulled the gate shut, it was standing like that at the gate.

The big male stopped. It turned and fled.

They were forced to make a decision about whether they were going to catch five or six, or just one.

And El Zermeño doesn't go in. I tell you that he had still another precaution to avoid danger. He didn't drink water at creeks. He only drank water in isolated puddles where nobody could get close to him. Very cautious. Little puddles here, little puddles there. To avoid drinking water any place that might be frequented by people. He'd look, and he'd come drink.

What's so unusual is that he could have become so wild, so savage, having been a tame calf here that he ate tortillas, he could be led by rope. It was unusual. And then to become so dangerous.

The steer must have felt rejected.

Well, I think that he resented that they had taken him out of his home, and he isolated himself. They took his mates. His herd, in other words. We have to understand. And he was deeply hurt. I say he was hurt because he never again sought to join any other company. The same with crossing a plain. If another herd of steers crossed in front of him, he followed behind. Let them pass first. And another thing, he wouldn't walk on the trail. He found his own way through the brush.

Him alone. He followed no one's path. Teófilo would say, there's one unusual thing about this animal. Its intelligence. It's smarter, because a steer surely has more intelligence because it has no sex. Makes you think. There was not that need like a cow or a bull to go out for the needs driven by nature. No, not a steer. It focused everything, with no temptations. It focused all its intelligence on its own security.

And others like this one that went alone, lost, or that couldn't be followed or caught. There were others like this one.

Wild steers? Oh, yes. There were many.

El Zermeño

As The Implacable Fire,
Was His Fury,
When Overcrowded

The Only Way Out,
To Live,
To Be Free.

—R.M. Beasley

Nowadays these, the rodeo cowboys, they are timed at seven seconds to throw a bull. When Andrés Villareal rode a steer, he didn't fall off because it was for his life. I have to laugh when I tell this story.

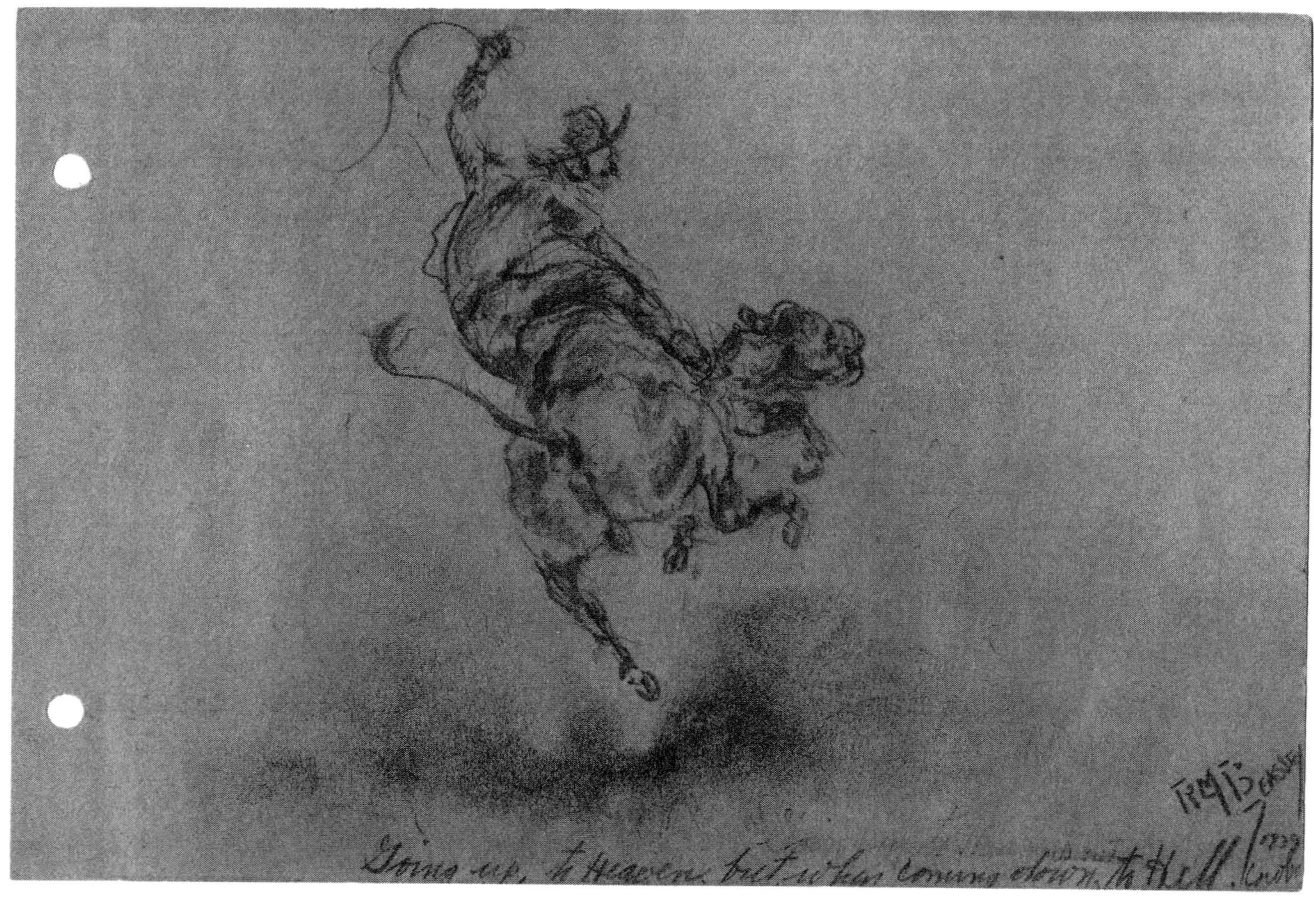

DRAWING 27
Riding a steer. *Courtesy of Jeanne Egbert Wright.* © R.M. Beasley Legacy LLC. Reprinted with permission.

Andrés Villareal

ESE ES ANDRÉS VILLAREAL, el que anda en el novillo.

Andaba en el arroyo, en la . . . pescando ladinos. Iban juntos campeando los dos cuando saltan dos novillos.

Este es Andrés Villareal. Ese era de los mejores que hubo aquí. No creas que este novillo también lo atacó.

Déjame decirte lo que pasó.

Cuando iban campeando él y Feliberto Garza, su amigo, se levantan dos novillos y cada quien le parte el de él.

Bueno, Feliberto lazó el de él pronto, y lo amarró y se fué a ver para donde había ganado Andrés cuando lo devisó que andaba en una loma, arriba del novillo.

Arriba del novillo.

Y el caballo corriendo porque lo habían corrido. Lo quería cornar, y corría el caballo.

"Y cuando llego allá," le dice que dijó Andrés, "Andale, hermano. Lázalo al novillo, que ya no aguanto las piernas."

Lo lazó Filiberto y lo estiraron. Dijó, "¿Como fuiste a dar arriba del novillo?"

Dijó, "Pues, lo lacé, lo tumbé, y cuando me apié a marrarlo, y ya para llegar, se me levantó." Dijó, "Si corría para el caballo, me alcanzaba. Si corría para el llano, podía reventar la reata y me mataba tambíen. Lo más seguro era arriba."

Lo montó. Ja ja.

Pero era que era muy jinete. Se atenía, ¿ves? Decía Filiberto que, "En cada vez que arrancaba el caballo, Andrés le tira al novillo de las manos, y lo tumbaba de costillas." Pero esta vez, como andaba en ancas aquí en la calda, no le hacía nada en las piernas de quebrarlos.

Y se levantaba otra vez el novillo.

Guapo.

THAT'S ANDRÉS VILLAREAL in the drawing, the one that's mounted on the steer.

He was in an arroyo, at the . . . rounding up wild steers. Both of them were going through the brush when two steers jumped out.

Andrés Villareal was one of the best ever here.

But don't think the steer attacked him in this case too.

Let me tell you what happened.

As he and his friend Feliberto Garza went into the brush, two steers came up, so each vaquero took off after his own.

Well, Feliberto lassoed his right away, so he tied it and went to see where Andrés had taken off to when he saw that he was up on a hill, mounted on a steer.

Riding a steer.

With his horse running off because it had been run away. The steer was trying to gore the horse, so the horse was running away.

"So when I get there," he says that Andrés told him, "Hurry, brother. Rope the steer, my legs are giving out."

Filiberto lassoed it and they tied it up. He said, "How did you end up riding a steer?"

He said, "Well, I lassoed it, threw it down, and when I got on foot to tie it down, just as I got there, it stood up on me." He said, "If I tried to run to my horse, it would catch me. If I ran for the open field, it would break the rope and kill me anyway. The safest thing was to jump up on it."

He rode it. Ha ha.

But only because he was a very good rider. He could hold on. That's the only reason he could do that.

Feliberto said that "every time the horse would take off, normally Andrés would throw his hands over the steer, and he'd throw it on its side." Bulldog it. But because he was on foot, out in the field, he could only bring it to its knees, and it would stand up again.

Talented.

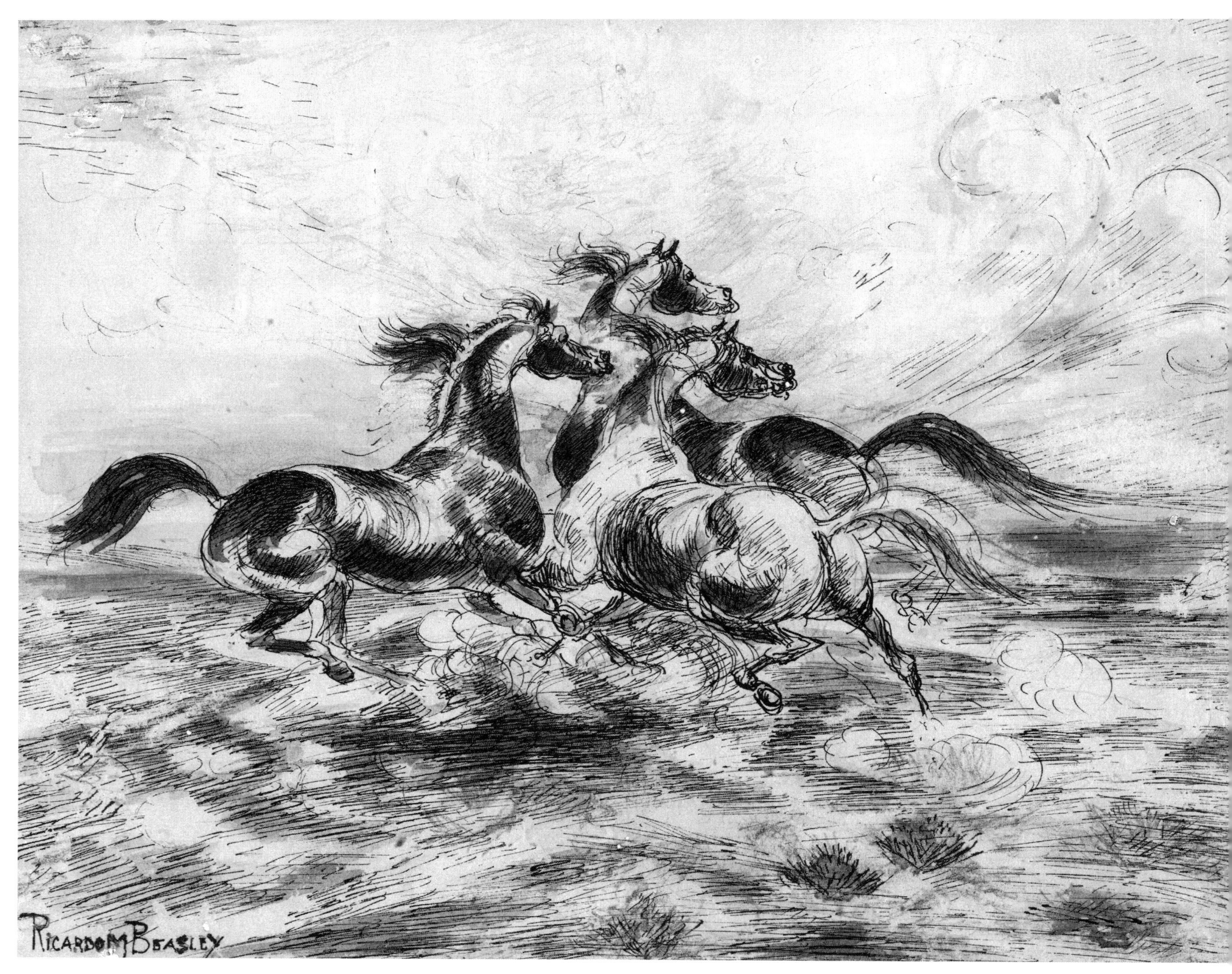

DRAWING 28
The mesteños. © R.M. Beasley Legacy LLC.
Reprinted with permission.

CHAPTER 4

The Mesteños

Editor's Note:

The official dictionary of Mexico defines mesteño as a wild horse in Texas.[*] Like the Texas longhorn, the mesteño (or mustang) was bred on the open ranges of the original Tejano land-grant ranches of South Texas. By the millions these wild horses propagated on the Tejano ranch lands and then dispersed across Texas and North America. And like the novillo ladino (wild steer), the mesteño stood as an equal to Beasley's vaqueros and a loyal "friend." Vaqueros anthropomorphized the mesteño, as they appreciated its strengths and heralded it mystical qualities. More than one of Beasley's vaqueros paid homage in his story to the loyalty and nobility of the mesteños. One of the most poignant of these is visible in the drawing of the vaqueros who slept out under the stars in *Waiting for the Wild Steers* (drawing 20). The sacrifice of the exhausted vaqueros sleeping on the ground evokes symphathy until one reads Beasley's description of their horses' condition. After dashing through the mesquite thorns and thickets of the chaparro negro, "they'd be covered with blood." Vaqueros often spoke of the

[*] Francisco J Santamaría, *Diccionario de Mejicanismos* (2nd ed.; Mexico City: Editorial Porrua, 1974), 719.

mesquite thorns that had become so deeply embedded in their chest or arms; they simply lived with them, much like a wounded veteran would live out his life with bits of shrapnel in his body. The old vaquero, Don Estevan Rodríguez, said that the horses in this drawing were so beaten that "they would not eat but would lay down and stretch out in pain, and you could hear the moaning of their suffering all night." Like the vaqueros, though, the horses leapt at the thrill of the chase: "the moment they heard or sensed the steers nearby, they became so excited over the next wild chase that none of them was lame anymore." They reportedly tugged at the reins to alert the sleeping vaqueros to the steers approaching in the dark. As a vaquero said, "They were like the Vaqueros themselves." One of the night chases depicts a vaquero riding full speed through the dark when his horse suddenly breaks to one side. Just as the vaquero complains at the sudden turn, he gets the whiff of musty dust from the steer that was charging at him in the dark. He was saved by the quick action of his loyal "friend."

The history of South Texas is genealogy to the Tejano. For example, one of Bealey's drawings depicts a majestic black mesteño stallion, standing in front of the ruins and a tombstone with the inscription "Santos Moreno QDEP" the Spanish equivalent to the English "RIP." The background story of this drawing is that Santos Moreno died before completing his dream construction project. Like the main character in Earl Hamner Jr.'s book *Spencer's Mountain*, he never built his dream chapel for his ranch family. Beasley embeds three themes in this one drawing. First, the ruins of the unfinished chapel. Second, the tribute paid by the wild mesteño stallion last seen standing at the ruins. And third, Santos Moreno was Ricardo M. Beasley's grandfather. As the grantee of the Spanish land grant La Trinidad, he was the father of Beasley's mother, Concepción Moreno—the "M" in Beasley's name. The drawing embodies the deep roots that these vaquero families had to their heritage and to their Spanish land grants. ★

THE WILD HORSE,
Symbol of freedom, of courage,
Of determination, and of pride,
Has been one of my greatest inspirations
Ever since childhood.
No one else has ever displayed more
The fire of that love than the Mesteño.

They have been an inspiration
Even to those men of great spirit.
That wild snort of defiance still seems to ring in my ears,
And that flying mane still is in my memory,
Blazing through the wind
Like a flag in shreds, but never conquered.

The Mesteño might have vanished,
But through no fault of his.
He did his best.
He disappeared far beyond horizons
Where men will never reach . . . but not into oblivion.

So great was his spirit, that he left behind a trail of memories
Never to be forgotten.
The Mesteño and those who speak of his glory,
Remind me of the monument to the fallen soldier,
Who, nearly dying, halfway raises on his elbow
And, with outstretched arm, hands the last shell to his comrade,
Still standing, but with an empty gun,
And he, in turn, replies with love,
"Sleep in peace, my dear brother;
The country asks for my hand to charge again."
—R.M. Beasley

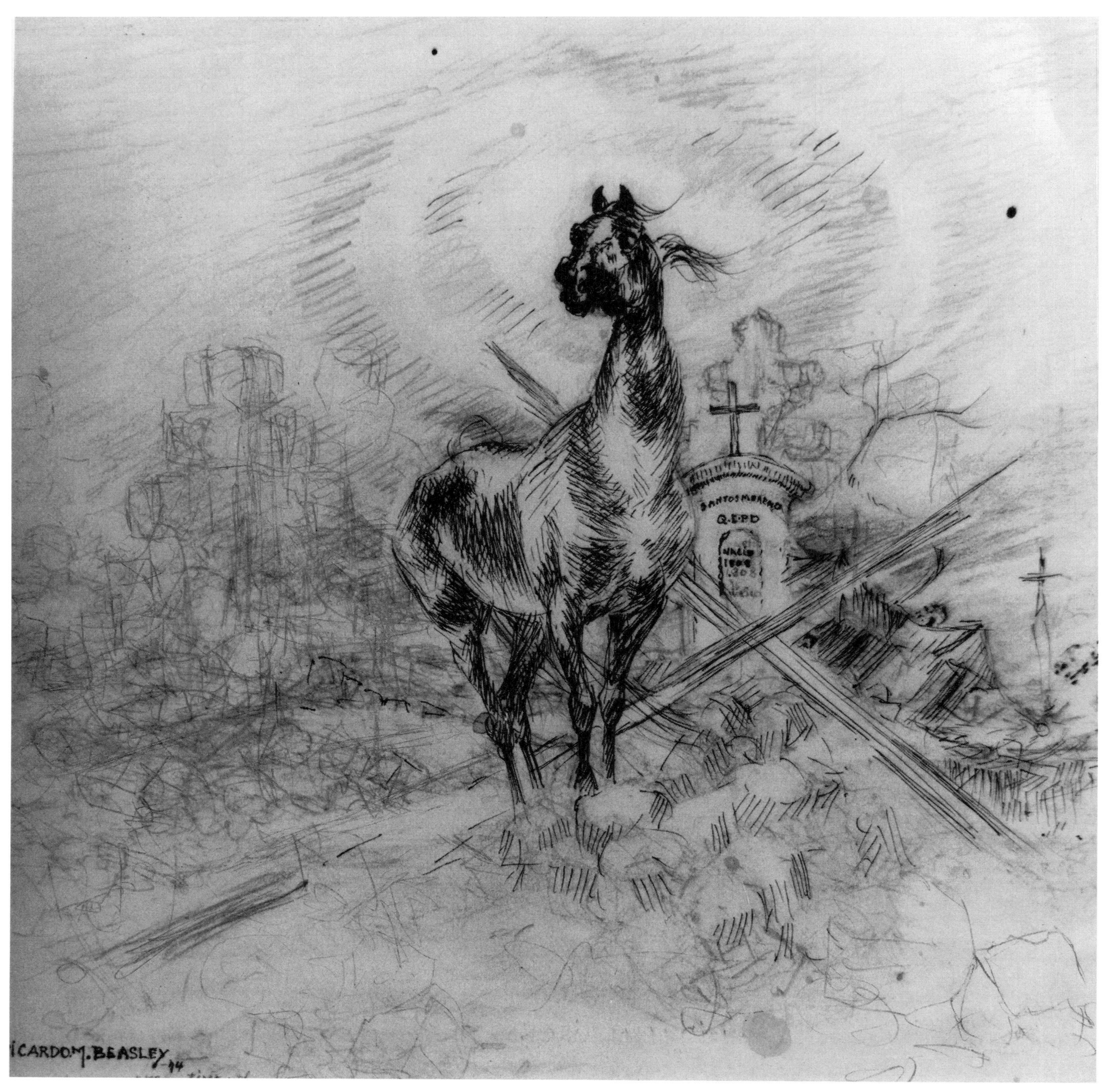

DRAWING 29
Rest in peace, Santos Moreno. © R.M. Beasley
Legacy LLC. Reprinted with permission.

Rest in Peace: Freedom Lives

Alone he stood in an aura of desolation,
Jabalín, the last of the mustang blood,
His country lost and he a captive.
A picture of misery, almost dry to the bones,
The sign of a long torture, of anguish,
But still defiant, a shadow of death, but in heart and spirit,
Fiercely alive as the last of the Indian braves,
Who before having his spirit crushed
Would fight to the death rather than exist without pride.

As by decree, he had to be where he stood,
By the catacombs: in the ruins of the unfinished church,
As if to pay tribute and bid alast goodbye to the remains
Of Santos Moreno, who had given a country
To his ancestors and him, to me and mine,
When this was still the land of the wilds of Mexico.

It must have been the course of destiny at the end of the story
Of the home ranch, La Trinidad,
That the two braves who stood alone,
One at the beginning, the other at the end.
They that never surrender in the face of horrors
Should come to be together as if in spirit only,
Before parting forever, as in honor,
Although in disaster because it was not defeat to them who couldn't help it.
If only they could have wept for the great love for the Land lost.
His fate now was that of an orphan
Or that of a child abandoned at the door of a convent,
Facing the act of some kind soul that would give him
A home and respect his pride, or the act of some cruel heart
That would destroy his life and a proud era in one stroke.

REQUIEM

As a flare in the ashes, a smile among the dead,
That is the great spirit of the brave, of the free,
High above the darkest hour, he rises,
Blazing the trail of the most sacred heritage.
It is a symbol in sunlight, as a crowning of the blessed,
It is an emblem of the heavenly light, in gold leaf,
The God given gift to stand free, and not to crumble in despair.

Ricardo Moreno Beasley
Wednesday, August 14, 1974
Alice, Texas

(© R.M. Beasley Legacy LLC. Reprinted with permission.)

The Vaquero's Horse

Their great cowhorses of Spanish origin add to the building of the character of the Vaquero, to their dignity. It was something to be very proud of, to be very dependable on his duty. The horses would try to kill themselves to comply with their master's demands.

That horse had a character of his own; he was as much on his own as the Vaquero himself. He had the same principles, the same heart, the same spirit, the same determination, and the pride to be self-reliant. Those great horses didn't need a whip or spur to do their best. They were like the Vaqueros themselves. The best was done, not for honors, but for the same urge of life itself, and for the will not to fail.

Those horses were something to envy for their courage, endurance, and determination to succeed. It took the best of the Vaqueros to stay with a horse of that kind in the brush country. Those horses excelled on their own so much that sometimes the Vaquero had something to learn from them. They gave the Vaquero a feeling of pride, of gallantry, to have them for a campanion, so dependable in time of trouble.

It is with these great cowhorses and the blessing of his Indian heritage that the Vaquero excelled above all. These gave the gift of nature for his guide, his wisdom, his knowledge of animal and human nature, his instinct, his courage, the love of the wild.

The horses and the Vaqueros would get pretty crippled sometimes. While running a wild steer through the thicket, horses would get beat up so bad in that infernal black brush Chaparro Negro, that they'd be covered with blood. It is hard to believe the extreme suffering that the horses reached after a hard day of riding through the brush, to what extreme was the fire of their spirit at the call of duty, even to ignore the anguish of such suffering.

An example of such torment, such a sense of duty was given to me by an old vaquero, Don Estevan Rodríguez. Don Estevan said that when they were roping wild steers on the Seeligson Ranch, after a hard day's ride, the horses would suffer so much that when they tied them after dark to graze them, they would not eat but would lay down and stretch out in pain, and you could hear the moaning of their suffering all night.

Admirable as it may seem, the next morning when they were out looking for the wild steers, the horses were still all crippled up, but the moment they heard or sensed the steers nearby, they became so excited over the next wild chase that none of them was lame anymore. Their love for the life of excitement was so great that they forgot all about themselves. The Vaqueros had this same love. They overlooked the suffering, the danger, for the conquest of the goal ahead, the wild steer.

© R.M. Beasley Legacy LLC. Reprinted with permission.

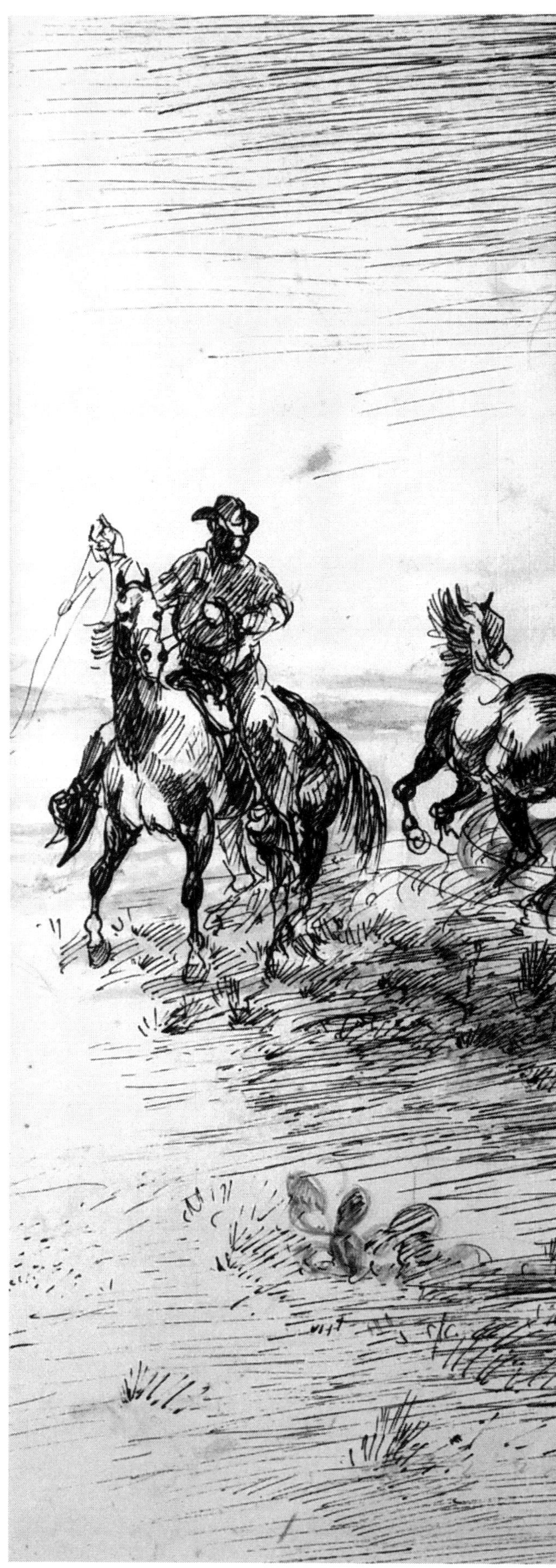

DRAWING 30
Sam Smithwick going after the mesteños. © R.M. Beasley Legacy LLC. Reprinted with permission.

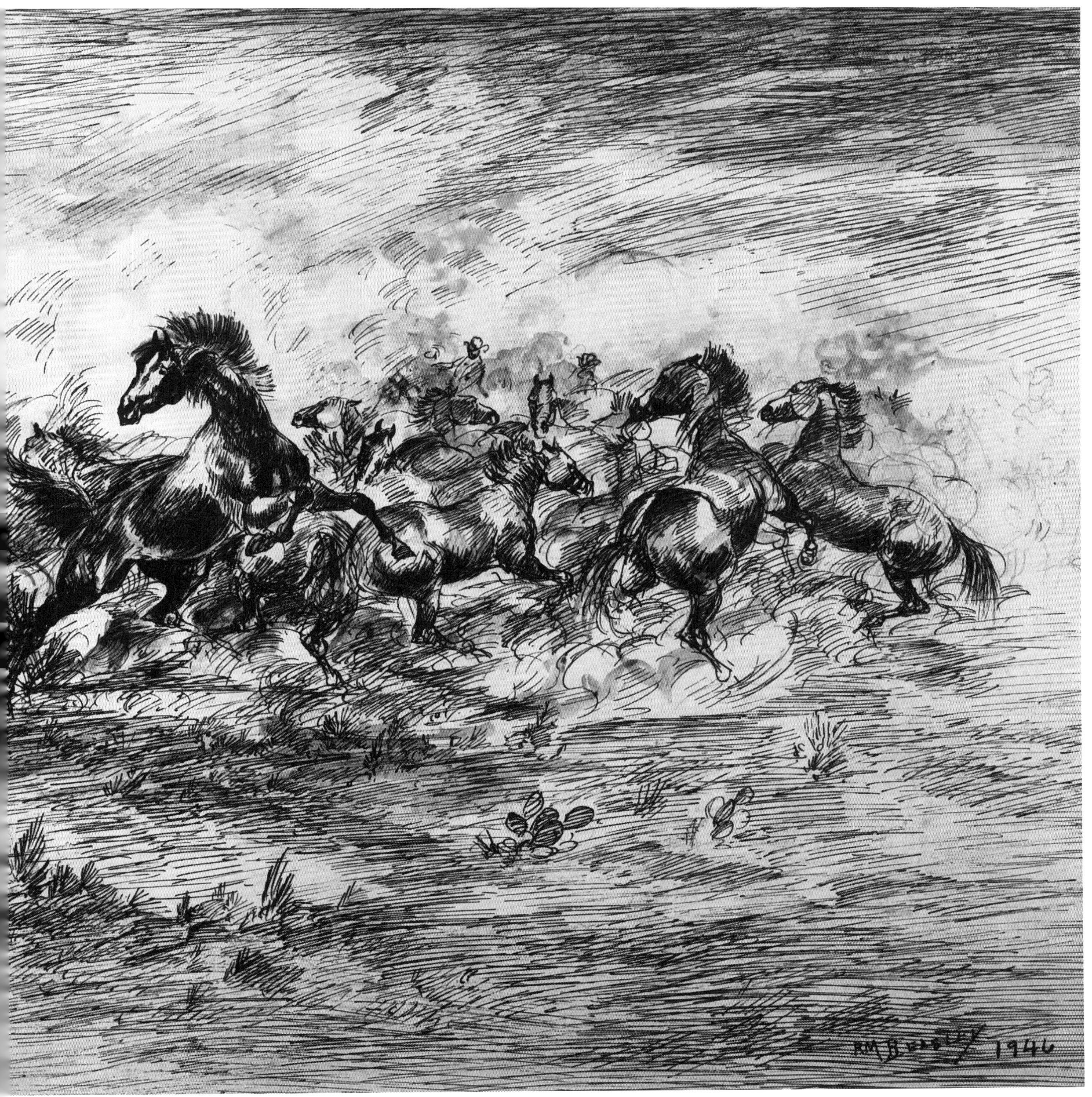

El Acordeón

ESE VAQUERO que va volando de la silla en ese retrato, ese es Juan Everett. Estaban lazando en, esa historia pasa en el rancho del Dobie en nineteen five. Y se habían ido unos, habían estado lazando caballos, él y Sam Smithwick. Estaban de compañeros, y le habían comprado una caballada de buena sangre, good breed horses a un vecino (drawing 30). Los estaban domando pa' la silla, y los estaban echando a la trampilla en la noche, chiquita. Por cerca, tenía al lado del sur un estero del río. Estero son pozos que se hacen largos, al lado del río. A slough. Son pozos muy hondos, esteros de agua. Tan ondos que los caballos tenían miedo pasar. Los usaban de cerca y no se iban los caballos.

Pero estos dos que había allí eran más renuentes, se pasaban. Se pasaron, y se juntaron con los mesteños. Ellos se hicieron los lideres de los mesteños porque eran más inteligentes, mejor sangre y más ligeros. Y ya conocían al hombre, y sabían como defenderse. Los dos estaban mesteño de atiro, brutos. Allí es donde estaban los jefes. Y anduvieron volando los caballos con los mesteños.

En esta ocasión, dijó el patrón, "A ver si pescamos solos a los caballos que andan con los mesteños."

Y la manera que lo hacían cuando querían pescar caballo ladina—el potrero muy largo como catorce o quince millas—sabían donde comían los mesteños, los caballos. Se llevaron una remuda. Remuda quiere decir, este, pués en inglés, no hay palabra pa' remuda. About 40 saddle horses. Y los pusieron en el east side pasture los pusieron con unos cuantos vaqueros que los cuidaran. Luego lienzan ese que iba para el west.

Allá comían los ladinos a trece o catorce millas. De allá los iban a levantar, y cada media milla, había un vaquero

THAT VAQUERO, the one flying out of the saddle in this drawing, that is Juan Everett. They were roping in, that story takes place at the Dobie Ranch in nineteen five. And some had left, they had been roping horses, he and Sam Smithwick. They were partnering, and they had bought a herd of good stock, good breed horses from a neighbor (drawing 30). They were breaking them for the saddle. They were putting them in a pen at night, very little, and for a fence on the south side they used a natural estuary. An estuary is a long pool that forms along the side of a creek or river. We have scattered deep pools, water sloughs. So deep that horses were afraid to cross. Vaqueros used them as a boundary because the horses could not escape.

But there were two mesteños there that were surlier. They escaped. And they escaped, to join the mustangs. And they became leaders of the mustangs because they were smarter, better stock, and quicker. They were already familiar with man and knew how to defend themselves.

Those two were completely mesteño, wild brutes. They were leaders out there. And those mesteños ran with the mustangs.

On that occasion, the owner said, "Let's see if we can catch those horses alone, that are running with the mustangs."

And the method they used to catch wild horses—a large horse pasture about fourteen or fifteen miles—they knew where the mustangs grazed, the horses, so they took a remuda. Remuda means, well in English, there is no word for remuda. About 40 saddle horses. So, they put them in the east side pasture, and put them in with a number of vaqueros to watch over them, and then on a line of riders out to the west.

Out there the wild ones grazed at about 13 or 14 miles

empuesto. Cuando los vaqueros los corrían hasta aquí, los pescaba el otro. La cuestión era no dejarlos descansar. Cansarlos primero. Correrlos.

Para cuando llegaban allá a donde estaban los cuarenta caballos mancitos, iban en trampa ya.

Dejame decirte lo que pasó. Esa tarde los enceraron todos—los pescaron todos con los caballos mancitos, y los mudieron a un potrero chiquito. Y era muy tarde, obscuro.

Y dijeron, "Vale más ir a sacarlos." Y en la mañana, que vinieron a sacarlos, dijó Dobie, "Se nos van a ir los caballos," dijó. "Allí en el río se nos van." Dijó, "Vale más poner a Juan y a Sam en el orilla de la cerca y se los dejamos ir pa' que los bajen de pasada."

Y pusieron a Juan y a Sam Smithwick.

Y cuando venían los caballos corriendo, estaba aquí la cerca y había así un sendero y había monte amontonado. Ramas.

Cuando ensentaron los caballos, Sam cortó muy pronto y se le fué por detrás, y no pudo hacer nada.

Y Juan sí lazó, pero el caballo no quiso brincar la rama, y dió un tiron de parados terrible. Lo voló, en el aire. Y le sacó la silla por la cabeza, el caballo. Pero cuando se levantó Juan, caió en la reata con la mano y el dedo gordo y el otro dedo, y le cortó la carne haste el hueso como un alambre. Y cuando se levantó, ya tenía la silla enruedada en las manos. Traía la silla enruedada en las manos a caballo. Y llega Sam y lo laza.

Le decían "El Acordeón" a ese caballo (drawing 31).

Era un caballo muy reparador. Le pregunto yo a Juan— porque los vaqueros son muy inteligentes pa' ponerle nombre a los caballos. Y en ese tiempo, Juan vivía conmigo.

Y yo le digo a Juan, "Mira Juan, yo no le he podido hallar porque le pusieron "Acordeón" al caballo. ¿En que manera se parecía un acordeón?"

Juan dice, "Como reparraba," dijó "Se puntaba."

Ja ja ja ja.

distant. They went to get them there, a vaquero placed every half mile. When the vaqueros chased them here, another would catch them. The point was to not let them rest.

Tire them out first. Run them.

By the time they arrived where the forty remuda tame horses were, they were already in the trap.

Let me tell you what happened. That afternoon they corralled all of them—they caught them along with the tame horses and moved them to a small pasture. It was late, dark.

They said, "We'd better get them out."

And in the morning, when they came to take them out, Dobie said, "The horses are going to escape," he said. "At the river, they're going to escape." He said, "We'd better put Juan and Sam at the edge of the fence and leave them there to block the way."

So, they set Juan and Sam Smithwick.

And when the horses came running, the fence was over here and over there was a path and there was a pile of brush. Branches.

When they set the horses, Sam quickly cut toward the rear, and could do nothing.

Juan did rope, but the horse balked at jumping the brush, and starting bucking terribly. It threw him. Off his horse. And the horse pulled the saddle off over his head. But when Juan stood up, he had fallen with his hand on the rope with the thumb and forefinger of one hand on the rope, and and rope cut his flesh to the bone like a wire. And as he stood up, his hand was still entangled in the saddle. He had his hands tangled in the saddle of the horse. So Sam arrived, and threw his rope.

They called that horse "The Accordion" (drawing 31).

It was a real bucking horse. I ask Juan, because vaqueros are very clever in naming their horses. And . . . at that time, Juan lived with me.

And so I say to Juan, "Look, Juan, I can't see why they named that horse "The Accordion." In what was was he like an accordion.

Juan says, "The way he bucked," he said, "He stood on his nose."

Ha ha ha ha.

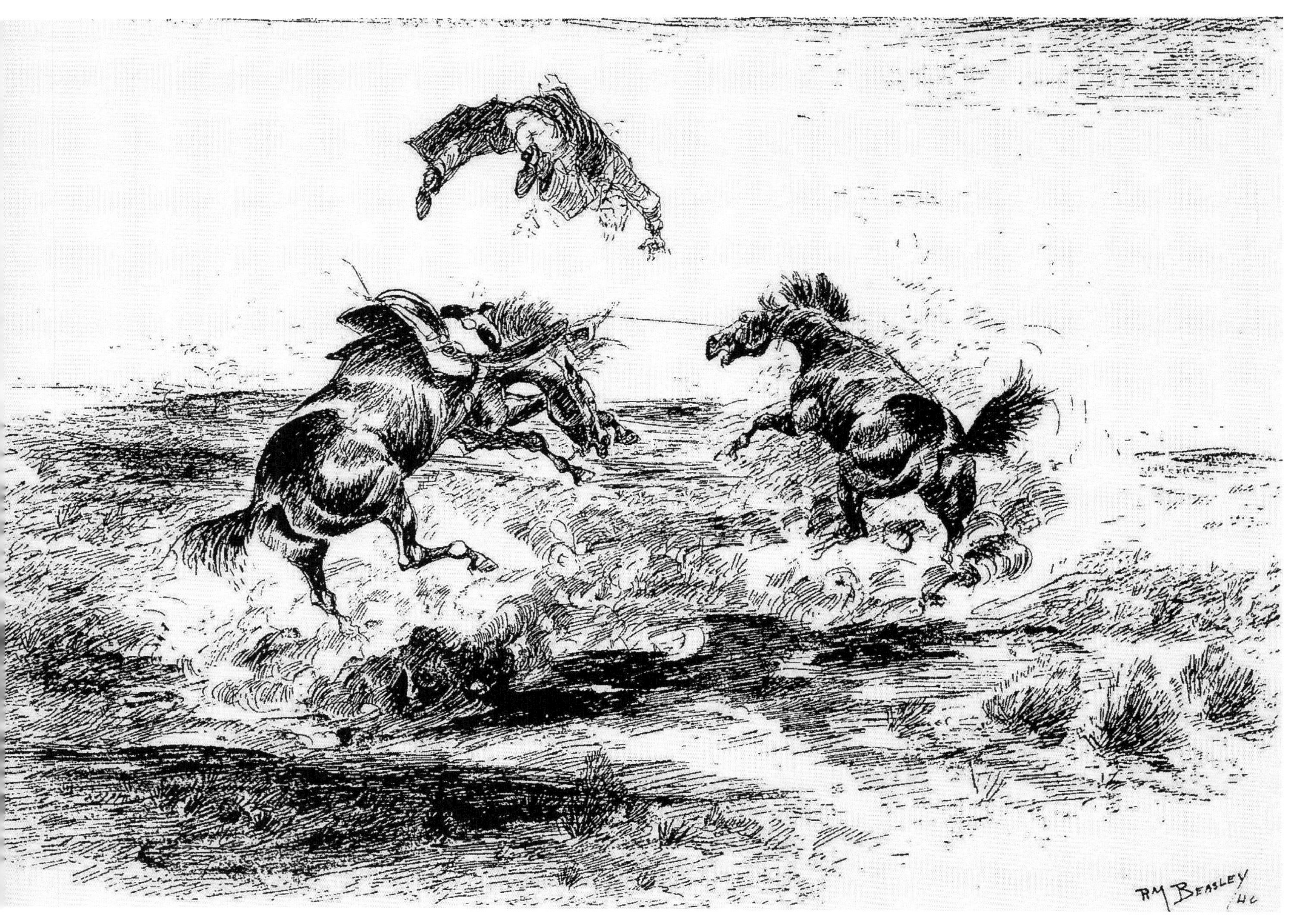

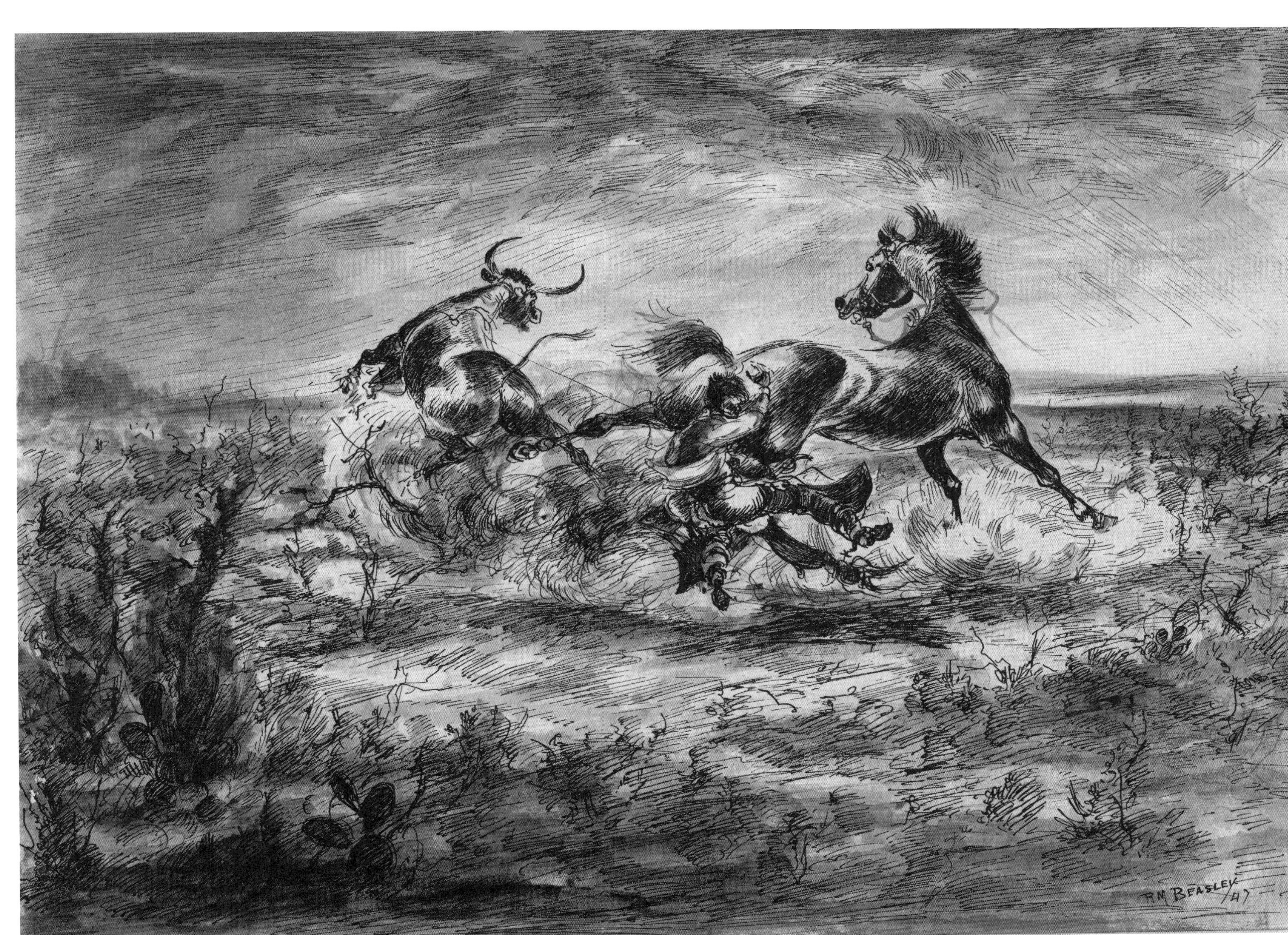

DRAWING 32
"El Títere Negro." © R.M. Beasley Legacy LLC.
Reprinted with permission.

"El Títere Negro"

AQUÍ EN DUVAL había ladinos caballos también—mustangs.

Había para, en los treintas or the late twenties, había una caballada ladina aquí. Dejame enseñarte.

This is the black stallion. Ese stallion fué todo lo que pescaron nomás en un laso. Los de más iban a matarlos todos con carabina. Arthur le dijó al hombre, los vaqueros . . . Arthur entró al potrero y le dijó, "Los pesca usted o me los vende y yo sé que hacer con ellos porque yo necesito el potrero."

Dijó el vaquero, "No te los vendo." Eran bravos, mataban mucho caballo. No había chanza de pescarlos.

Nadie les ponía mano.

Pero yo no sé si sería cría de mesteño aquel pero eran ladinos porque es mucha diferencia de un mesteño a la parada ladino. Ladino cualquier animal puede ser ladino, pero el mesteño is a breed.

Pero esos eran ladinos. Y ese caballo prieto lo pusieron a la silla. Fíjate que tenemos un muchacho aquí que parece indio. Sánchez, Hernández. Lo conoce mucha gente.

Trabajaba allí en la yarda del Community Center. Prieto, él. Puro indio. Y trabajaba en ese tiempo con los Spohn.

Bueno, Chito Muralla era el caporal. Y lo traía en la silla, nomás que nunca fué el cerro verde. Todo el tiempo era potro verde.

Y un día que estaba lloviendo, estaban hechando ganado al corral. Se perdió una vaquilla y fué atajarla este Chito en ese caballo. Y donde la vaquilla se va para atrás, el caballo se le pone en las patas. Y aquel dobló la pierna para no caer de bajo. Y se haría de como estaba él mojado, se había

HERE IN DUVAL there were wild horses—mustangs.

There were by, by the thirties or the late twenties, there was a wild herd here. Let me show you.

This black stallion was the only one they caught by lasso-ing. The others were going to be killed by shooting them.

Arthur told the man, the vaquero . . . Arthur went into the pasture and told him, "You either catch them or sell them to me because I know what to do with them because I need this pasture."

The vaquero said, "I won't sell them to you." They were wild, they killed a lot of horses. There was no chance of catching them.

Nobody could touch them.

I don't know if these were that mustang stock, but they were wild because there's a big difference between a mustang and a wild horse. Wild—any animal can turn wild, but the mustang is a breed.

But these are wild. And they put a saddle on that black horse. You know we have a young man here that looks Indian. Sánchez, Hernández. He's pretty well-known. He used to work over at the Community Center grounds. He's very dark. Pure Indian. Back then, he worked for the Spohn. Chito Muralla.

Well, Chito was the caporal. He was in the saddle, but he never was the big tower of strength. He was only a greenhorn.

So, one rainy day, they were putting a herd in the cor-ral. One of the heifers got away, and he went out to rope it on that horse. And when the heifer backed up, his horse rears up on its hind hooves. Chito bent his leg back to avoid

destrobado la media espuela. Se le ganchó el estribo, y como estaba mojada la vaqueta, se le enrededó en la pierna. Y se levantó con él caballo reparando. Reparó que no lo tocaba el suelo, y luego se avocanó el caballo y se lo hecho por entre las piernas. Y este se puso—humor del caballo ladino, muy coludo—se pescó de la cola y se levantó para arriba y le pegaba al capazón en el pecho, y sin saber para donde iba el caballo.

Y Chito le dijó a los vaqueros, "No lo corran, porque al cabo no lo alcanzan. Déjenlo a ver si busca la remuda."

Y se corrió por una labor, lo montó Simón Chico. Y al rato volvió con aquel se había de bajo, y lo pescaron cuando se pone entre los caballos.

Holding on for dear life.

Y para bajarlo de donde estaba enredado, le levantaron una pata al caballo y lo tumbaron con todo y silla. Le tumbaron la silla al caballo. Y la razón que nunca le pegó una patada era que el caballo estaba zambo y tiraba las patadas para afuera. ¿Entiendes? Eso lo salvó. Si no, lo mata el caballo.

Ni con, ni un . . .

Pero fué tanto el susto de él. Había un arroyo que estaba allí cerca, que cuando lo soltaron, arrancó y pasó el arroyo como un loco. Pobrecito. Y se le volvió un susto, una enfermedad, que ya es nomás huesos y pellejo.

Había una curandera allí en La Bandera, María Chahin. Fué con ella y ella lo curó de susto.

Me dice, "Archie [Parr] pagó la cura, cuarenta pesos." Y se salvó.

Hay está todavía. Pero imagínate que susto no será ir debajo de un potro, tu, sin saber para donde vas. Imagínate como se va sentir un hombre, tu.

Le decían al caballo "El Títere."

"Títere Negro."

sliding under. And perhaps because the leather was wet, the half-sole of his boot separated, and he got tangled in the stirrup. Then the horse stood up and started bucking.

It bucked so that its hooves never touched the ground.

Then the horse took off running, and he rolled under between its legs. So, this guy gets—a typical wild horse, it had a long tail—he reaches back and grabs a hold of the horse's tail just as it reared up again, him banging against the gelding on its chest. No idea where the horse was running.

So Chito told the vaqueros, "Don't chase it, because you won't catch it anyway. Leave it alone. Let's see if it looks for the remuda."

So, it ran off through a field, and Simón Chico mounted it. And after a while, he came back with that guy hanging on underneath, and they were able to catch it when it went in among the other horses.

Holding on for dear life.

Well, to get him out of his tangled hold, they had to lift the horse's hoof and throw it, saddle and all. They pulled the saddle off the horse. And the reason the horse never kicked him underneath is because the the horse was bow-legged and kicked outwards. Outwards. That saved him. Otherwise, the horse would have kicked him to death.

Not even with, not once. . .

But he got such a fright. There was an arroyo over there nearby, and when they untangled him, he took off running like a mad man. Poor guy. In fact, the shock never left him, an illness, that he was left just skin and bones.

There used to be a curandera over at La Bandera, María Chahin. He went to her and she healed him of his fright.

He says to me one day, "Archie [Parr] paid for my treatment, forty dollars." And he was saved.

He's still there. But imagine what a fright it must have been, going underneath a colt, not knowing where you're headed. Now, you imagine how a man is going to feel. They called that horse "The Puppet."

"The Black Puppet."

Victor Suarez and the Bronco

In the battle for freedom, and the honor of duty,
Death comes in between,
But it is worth it, as if a code to live by,
After all, what is life without freedom, without sense of duty?
A failure, not worth living, unless the struggle lives on
Or the existense of the two highest values of life.
Freedom and duty.

Both the bronco and the Vaquero, Victor Suarez,
Were right in their battle, each one did his best.
Victor, to die rather than surrender,
The rebel bronco, to kill to be free.

Victor didn't fail to do his duty, he was a great rider,
But he was a victim of an unexpected consequence,
His bronco shot out of the saddle over its head
With Victor along, as if the wish of the bronco
Had come true in his thirsty search for revenge,

For a moment death lingered over the Vaquero
As the wild horse crushed his chest in bawls of rage,
The echo of death in the country side.

Victor lived though. The vaquero with him saved his life,
but to him it was not enough to be alive and defeated,
He rode the bronco again, right then,
And he won.
The bronco gave up the battle.
—R.M. Beasley

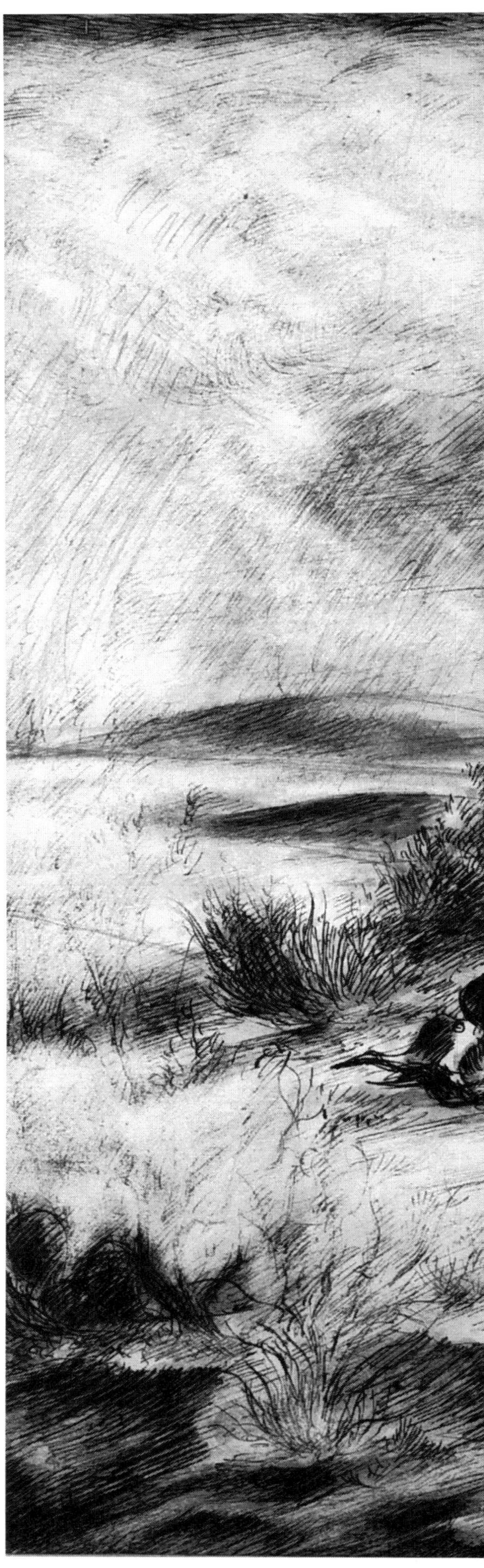

DRAWING 33
Victor Suarez and the bronco.
© R.M. Beasley Legacy LLC.
Reprinted with permission.

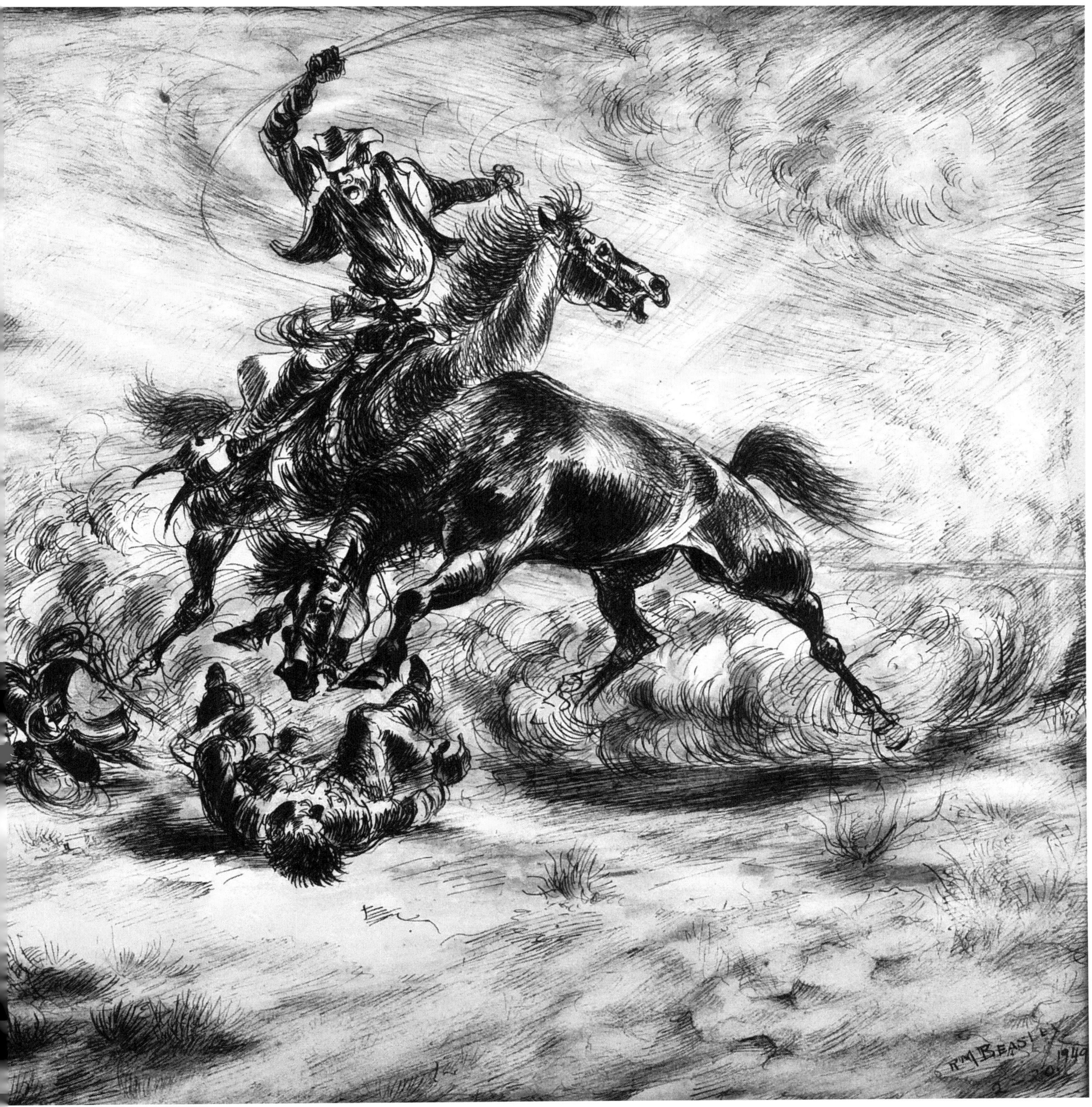

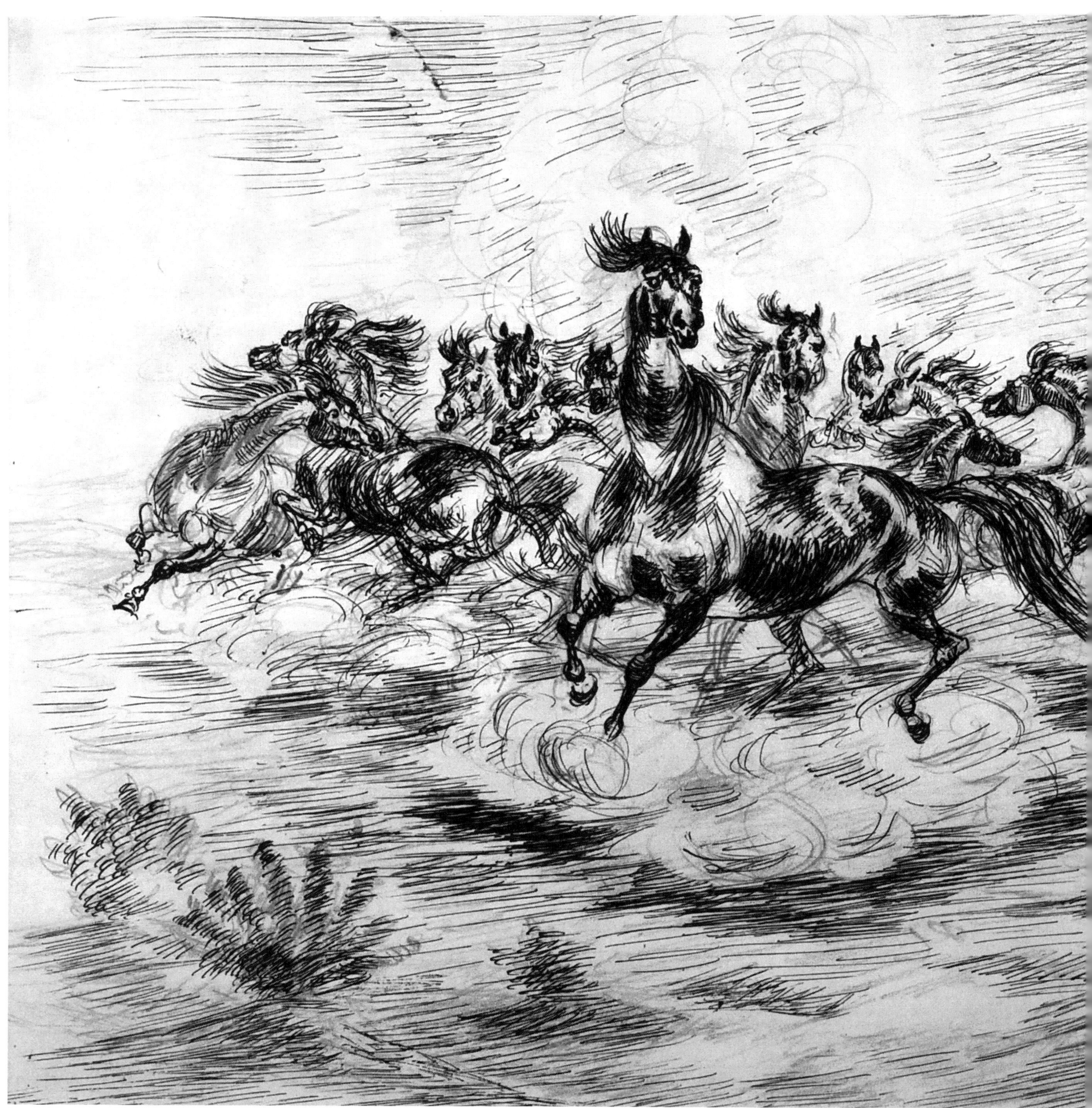

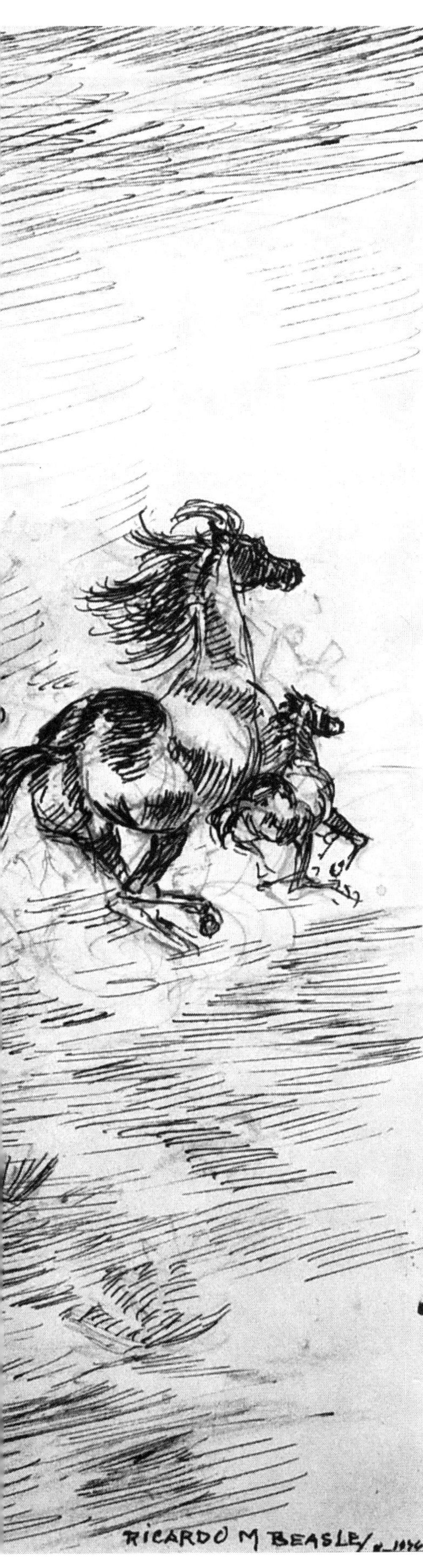

Supreme

If ever Love and Freedom nested together, supreme,
It was never greater than in the heart of that gallant,
Mesteño Stallion in the Amargoso Range,
During the time of the agony of their existance.

Heroic, he was, if anyone deserves a monument to his glory,
It is he, and those who in battle,
Sacrifice their lives for a great love and the cause of liberty.

So great was his love, so defiant his courage, that
At the first sight of the Vaqueros,
He would chase away his mares and offspring to safety.

And then come back to challenge them,
Close enough to encourage a chase,
As if an easy conquest, then would fly away,
To mislead them far from those he loved.

He must have felt as sure of himself
As the Eagle that soars the skies.
That stallion was never a captive for a moment,

But one day Fate failed him,
He killed himself
When he stepped in a hole and broke his leg,
He was lost into a world of his own.

DRAWING 34
Supreme mesteño stallion.
© R.M. Beasley Legacy LLC.
Reprinted with permission.

To the Vaqueros, a vain conquest,
His remains only, never his spirit,
Even at death he had won.
Those he loved, what he stood for, remained free forever,
All that was ever seen of them, was a cloud of dust

Vanishing on those horizons, without confines,
As was their freedom,
The heritage of that gallant, leader.
—R.M. Beasley

Only Death

Amazing was the courage
And determination of some of the horses
Of the Vaqueros in this brush country.

Only death or a near mortal injury
Could stay their aim:
The roping of the wild steer.
Neither man nor steed knew fear of danger
Or consequence,
So great was their spell.

Strange, the power of that force of will,
It was like that of a blind sacrifice,
Sometimes to the death,
But as with a Vaquero,
Without complaint of the fate,
So glorious
Was the feeling of both lives.
—R.M. Beasley

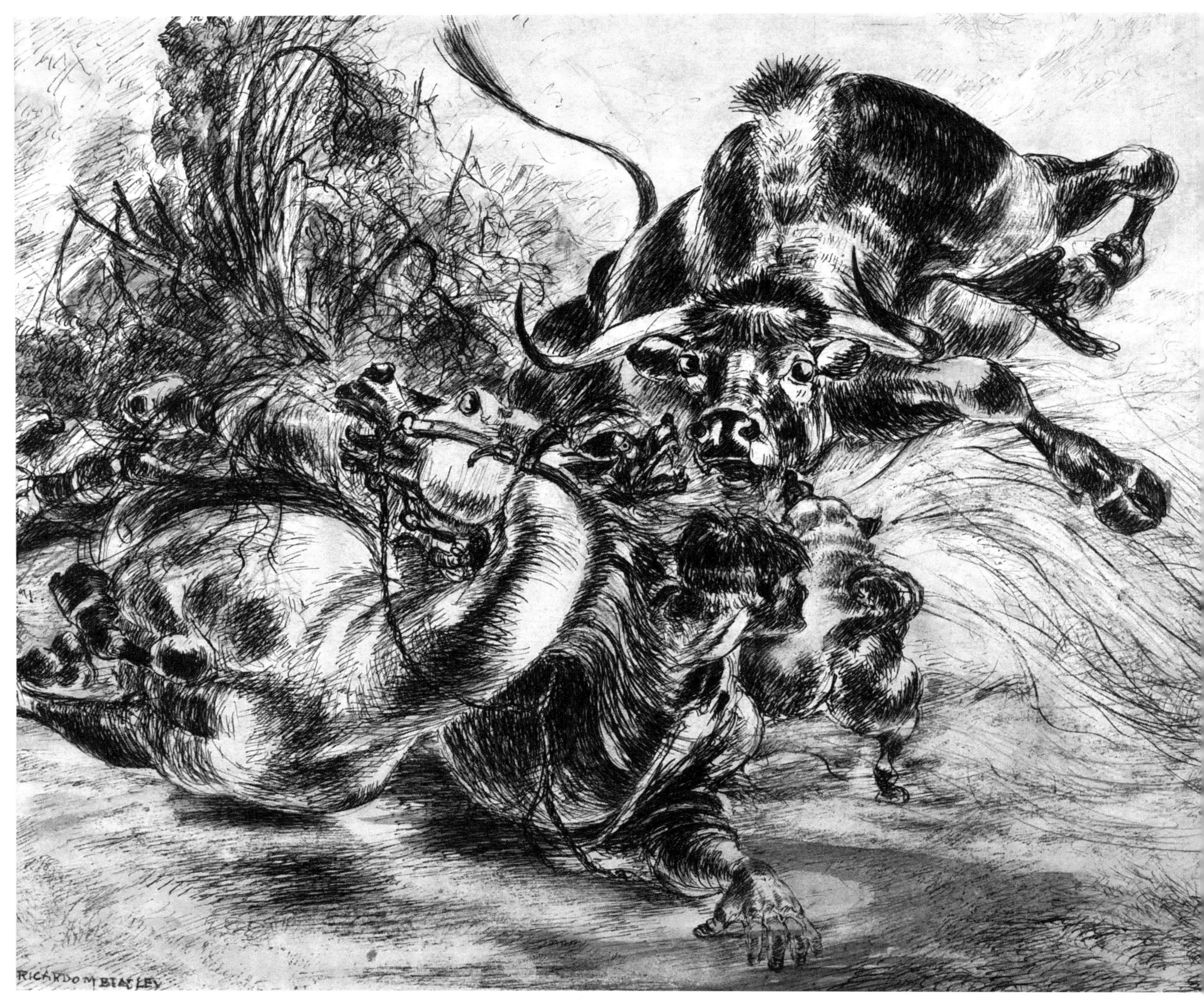

RICARDO M BEASLEY

His Boot Was Stuck

ESE FUÉ EN TAMBIÉN LA GLORIA.

Iba Feliberto Garza y ese otro vaquero. Iban juntos cuando saltaron dos novillos. Y cada quien se fué con el de él, y Feliberto dice que el de él se caió con un rama, y lo pescó caido.

Feliberto era el caporal allí en La Gloria.

Y luego que amarró su novillo de él, dice, "Me salí a buscar al compañero." Y dice, "Me puse a escuchar, y oyí que gritaba muy lejos."

Pero dijó, "No puede ser tan lejos . . . estar tan legos." Dice, "Ahorita nos arrancamos." Y fué y vió, dijó, "O está muy lejos, o está gritando dentro de un pozo."

Fíjate del conocimiento.

"Entonces me fuí a ver. Al poco, le dí de muy cerquita."

Fué cuando lo vió que sacó el pié de la bolsa, y andaba el caballo—el novillo se bajó a cornar el caballo, y él andaba arriba corriendo, que no sabía qué hacer cuando llega Feliberto. Andaba descalzo con una bota nomás.

Y dijó, "Lasa el novillo," dijó.

"Me mata mi caballo."

Y ya le había quitado el freno al novillo. Tocó la coincidencia que si no saca el pié, de seguramente que la bota estaba pescada de, en lo macizo. Si no, no saca el pié. Si no se sale, lo mata el novillo cuando se va.

Pues, este incidente pasó, I would say en los veintes.

En este dibujo puedes ver el terror hasta del caballo. El vaquero, el vaquero está preparandose a escapar.

Pero el caballo no se puede levantar porque caió en covacha. Y cuando está en covacha, con la altura, no se puede pescarse. No se levanta.

Y el novillo listo para atacar.

El novillo andaba buscando por donde bajarse.

THAT ALSO HAPPENED IN LA GLORIA.

That man Feliberto Garza was going along with another vaquero. They were riding together when two steers jumped out. Well, each one went for his own steer, but Feliberto says that his steer tripped on a branch, so he caught it on the ground.

Feliberto was the caporal over at La Gloria.

So, after he roped and tied his steer, he says, "I went out to look for my companion." And he says, "I listened for him, and I could hear him shouting very far away." But he said, "It can't be him so far away . . . so far away." He says, "We just now separated." So he went to look, and said, "He's either real far away, or he's shouting in a pit."

This shows how smart he was.

"Then I went to look. Right away, I came up on him very close by."

That's when Feliberto found him standing barefoot up top of a gully, with only one boot. His horse was trapped in the gully, and the steer was stooping down to gore the horse. The horse had fallen on his leg, and his boot had been caught snugly under solid weight of the horse. He had managed to slip his foot out of the boot.

The way it happened was that if his foot hadn't slipped out—because his boot was stuck—he would not have slipped out. If he doesn't get out, the steer will kill him and the horse.

He had already taken the rope off the steer, and was at a loss for what to do, when Feliberto comes up. He said, "Rope the steer," he said. "It'll kill my horse."

This incident happened, well, I would say in the twenties.

In this drawing, you can see the terror even in the horse's eyes. The vaquero is still trying to escape.

But the horse can't get up because it fell in a trench. And when a horse falls in a trench, with a high bank, it can't pull itself up. It can't get up.

The steer is ready to attack.

The steer was trying to find a way down.

El Comanche

LOS CABALLOS SON MUY INTELIGENTES. Los vaqueros conocían sus caballos. Y los vaqueros tenían que ir preparados pa' ir a caballo en el monte duro. Los vaqueros iban con los pantalones de lona y con la camisa de lona. Ahorita te enseño.

Hay en ese rancho, hicimos una vez una travesura con ese pobre agricultor. Este Harry Skidmore, casi todos lo conocían a Harry Skidmore.

Era muy, bueno, el papá de él te dice que al él, lo trabajaban en la labor, y luego lo ocupaban también para que les ayudara con tal vaca.

Y me platicaba Willie, el hermano de Harry, que, dice "Este Harry es muy desgraciado. Le dí un caballo que le decían El Comanche, caballo de los vinateros que habían dejado en rancho. Y parecía un tonto viejo que no hacía nada."

Ja ja.

Dice Willie, "Andabamos arriando un ganado, y el primer monte duro que pasamos, se perdió una vaquilla. Y que le entra El Comanche, y luego este Harry no sabía lo que traía. No lo pudo domar por nada. Para cuando Salió con la vaquilla—sacó la vaquilla el caballo. Aquel Salió encuerado."

Ja ja.

Pues no iba preparado.

Pues no cuando, no creía que aquel caballo podía hacerlo todo.

Aquel no iba preparado entonces. No iba preparado por el peligro.

¡Ah, qué bárbaro!

THE HORSES WERE VERY INTELLIGENT. Vaqueros knew their horses. And the vaqueros had to go prepared for the dense chaparro. Those vaqueros wore canvas pants with a canvas shirt. I'll show you in a minute.

Over at that ranch, we once made a complete mess for that poor farmer. This Harry Skidmore, almost everybody knew Harry Skidmore.

He was very, well, his father tells you that they worked him in the field, and then they would assign him to help them with a cow.

Harry's brother, Willie, used to tell me that, he says, "That Harry is disgraceful. I gave him a horse that they call El Comanche, one of those vintner horses that had been left at their ranch. He was one of those crazy old men that never does anything."

Ha ha.

Willie says, "Well, we were driving a herd and lost a heifer in the first thicket we passed. So, El Comanche bolts into the brush, and Harry had no idea what he was riding. He couldn't control that horse for anything. By the time it came out with the heifer—and the horse did bring out the heifer, the rider was stark naked.

Ha ha.

He wasn't prepared.

Well, he didn't expect that that horse could do it all for him.

So that one didn't prepare.

No. He wasn't prepared for the danger.

Oh, that was crazy!

CHAPTER 5

Burros y Mulas

Editor's Note

For Beasley's vaqueros, even mules and donkeys asserted their pride and dignity. Beasley cites the day that his best friend, Juan Everett, was guiding the chuckwagon through the Nueces River bottoms, the dense underbrush scraping tightly against both sides of the wagon. The vaqueros were lulled by the humid, muffled air. Everett was looking forward to getting married at the end of that drive. The cook wrapped the chuckwagon reins around his knees and rolled a cigarette. The caporal allowed that they might camp down just ahead. The calm of the day was suddenly shattered by a *burro manadero* that came up from behind, trying to squeeze through the thick brush to pass the chuckwagon. Beasley explains that a burro manadero is a guide donkey that the vaqueros tied neck-to-neck with an unruly steer to weigh it down. But the burro manadero had to be aggressive and strong. In his taped interview, Beasley laughingly exclaims, "Brutes. Ha ha." The donkey startled and stampeded into a tree, crashing the wagon and the entire calm of the day.

This last chapter of Beasley's narratives ends with the winsome tale of an "Invisible Mule." One day, Juan Everett reported to the caporal that he had spotted mule tracks in the horse pasture. The caporal ordered him to drive the mule out of the horse pasture before it disrupted the grazing operations. But Juan had never actually seen the mule—only its tracks. The tracks were mixed among the tracks of steers, prompting Juan to wonder about how lonely it must be, like a man without a friend. Juan never saw the mule, and serenely resigned himself to the mystery of the "invisible" mule that he tracked for weeks, but never saw. Beasley recounts the old vaquero's wisdom as a metaphor for life: "It was very vast, the open range . . . and so very winding, the trail through horse pastures." ★

El Burro Manadero

ESTE ES JUAN EVERETT. Cuando se iba a casar, cuando, en novecientos ocho, cuando yo nací. Fueron pa' a traer un ganado de aquí de Duval, pa' el Río Nueces. Y acabandose la corrida que venían con el ganado, Juan se iba a casar.

Dice que se fué de aqui en la mañana, llevaba su carro de campo—el chuck wagon, esos vaqueros sus mulas y todo pa' Loma Blanca. Loma Blanca está al Río Nueces.

Bueno, ya obscureciendo, iban llegando, iban por el bottom del río donde había mucha arena. Era un monte muy duro que apenas cabía el guayín, el chuck wagon.

Y dice el remudero, el caporal, dice, "Nos vamos ir adelante pa' ir hacer lumbre," y dice, "No, que ya estaba el campo muy cercas."

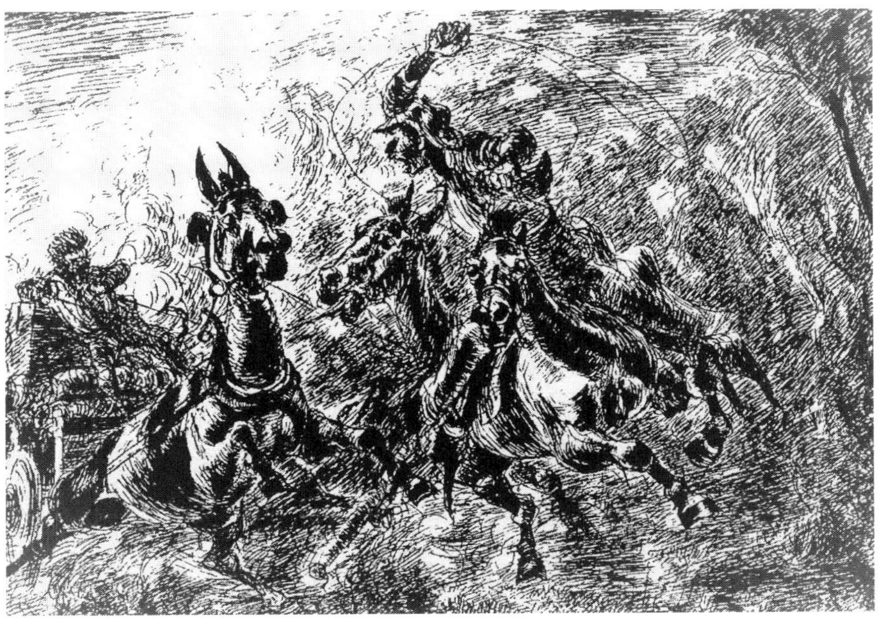

THIS DRAWING IS JUAN EVERETT. When he was going to get married in 1908, the year I was born. They were going to drive a herd from Duval to the Nueces River. And at the end of the drive, Juan was going to get married. So, he says they left here in the morning.

They had the chuck wagon, those vaqueros, their mules and everything to drive the chuckwagon to Loma Blanca. Loma Blanca is on the Nueces River.

At dusk, they were arriving, going along the river bottom where there was a lot of sand. It was a dense brush, hardly wide enough for the chuckwagon.

So, the *remudero*—the caporal—says, "We're going on ahead to make a campfire," and then he says, "No, never mind, the brush is too confined."

Well, the cook was driving the chuckwagon and decided to roll a cigarette while they were driving on the trail. He wrapped the reins around his knees, and he was rolling the cigarette while Juan—the guide—rode his horse alongside and led the chuckwagon and the mules. And as I said, the brush was very dense.

When suddenly, they hear behind them the braying of a guide donkey "(burro manadero)." And a guide donkey is a combative demon. A burro manadero was a donkey that they tied tightly to the neck of an aggressive wild steer to hold it back, keep it from bucking.

The burro manadero is not a wild animal. They're domesticated, but very aggressive brutes. The mules were frightened because the guide donkeys batter them a lot. When the donkey brayed, they lost control of the mules.

Juan says, "If I hadn't been working such a fine horse, they would have killed me and my horse."

Y el cocinero se ponía hacer un cigarro mientras manejaba el chuck wagon, y iban caminando. Se amarró las riendas en la rodilla, y iba haciendo un cigarro y Juan—el guía—iba adelante del guayín, de las mulas. Y como te digo, el monte era muy duro.

Cuando aquí de repente atrás—el rebuz de un burro maradero. Y un burro maradero es un díablo de bravo. El burro manadero lo amarraban al pescueso de un novillo ladino para detenerlo, amansarlo. Era mancito pero son muy descracíados, muy bravos. Las mulas tenían miedo porque las golpean mucho. Al rebuznar del burro, dieron control de las mulas.

Y Juan dice, "Si no trabajo tan buen caballo, me matan con todo y caballo." Y el camino muy angostito. El burro los venía sigiendo, pero no se podía pasar porque el monte estaba muy duro y el tierregal se lo estaba ahogando.

Dice, "Y yo saque la reata y les daba a las mulas en la cabeza."

Y luego se le soltó un zafadero a uno de los mulas. Traía el enjaecía en rastra.

Dice, "En una vuelta del camino, se arrancaron derecho las mulas a todo vuelo pa' el monte. Y fueron y pescaron un palo negro y calleron muy hechados y todo tan pronto calló muerto.

Y el Juan Everett. Él se iba a casar volviendo de allá. Iban a traer quinientos novillos de aquí de San Diego.

Ya estaba prometido. Y alcanzó a casarse. Se casó. El primer hijo que tuvo fué el papá de Concha.

Decía Juan, "Un día de estos le voy a dar un bruto de susto." Ja ja.

Del susto que llevaron ese día.

The trail was very narrow through the thick brush. The donkey was following right behind them, but he couldn't pass them because of such thick brush and choking dust.

He says, "So I took out my rope and whipped the mules on the head." Then the yoke slipped off one of the mules, dragging the harness.

Juan says, "At the turn of the road, the mules stampeded into the brush. And they went and ran straight into a large palo negro bush, and squatted, completely hunkered down, and it all came to a quick halt.

Ha ha ha.

Juan Everett was going to get married on his return from there. They were taking five hundred steers from here in San Diego.

He was betrothed, and he got married. The first son he had was Concha's father.

Juan used to say, "One of these days I'm going to give him a heck of a scare." Ha ha.

Like the fright they had that day.

La Mula Invisible

UNA NOCHE temprano estavamos platicando a la orilla de la lumbre del campo tomando café y la conversación era del tiempo de los novillos ladinos. Acavando yo de platicarle de un novillo ladino Kineño, muy peligroso, que abrían matado en la presa de La Carabina, me dice Juan, "de esa presa seguí yo una mula que nunca pude alcansar. Te voy a platicar como di con esa fuea [huella]."

Fué en la caida del año esa ocasión. Hacía poco que habían echado unos novillos en ese Rancho La Gloria y el trabajo mio era voltear los potreras y reconocer los novillos aver si no estavan gusanientos de las marcas por estar recién marcados. Además iba voltear las sercas porque un ganado en terreno nuevo les da mucho por agilarse por las sercas buscando salida. En veces se arriconan en una esquina por varios días y se encalman y se maltratan mucho. Y como no conocen el terreno, no ven donde estan los aguages, los papalotes, las presas o charcos. Hay que desparramarlos y arríarlos, enseñarles onde esta el agua. Lo mismo que por varíos días andan de monton juntos en varias partes del potrero. Hay que desparramarlos pa' que se hallan agarrando sus comederos que vayan reconociendo el potrero. Hay que ir a visitar los aguages segun va pasando el tiempo a ver más o menos cuantos novillos estan bajando al agua, ya bien sea en este papalote o el otro o en una presa aquí y otra más adelante para estar uno seguro que ya se estan haciendo los novillos al terreno, que no se maltraten.

En una de esas vueltas a los potreros de los novillos, llegue una tarde hay muy tarde a la presa de La Carabina a tomar aqua mi caballo y yo. Luego me fuí andando al paso por toda la orilla de la presa reconociendo a ver cuantos

EARLY ONE NIGHT, we were chatting at the side of the campfire, drinking coffee and the conversation turned to the days of the wild steers, and I, having just finished telling about one wild King Ranch steer—very treacherous—that they had killed at the dam of La Carabina (The Rifle), Juan says to me, "From that dam, I pursued a mula that I could never catch. I'm going to tell you how I came upon its trail."

That occasion was in the fall of the year. They had recently put one thousand steers in the La Gloria Ranch, and my work was to inspect the colts and the steers to see if they had worms where they had been recently branded. My job was also to inspect the fences because when a herd is in unfamiliar terrain, they spend their time probing the fences in search of a way out, and sometimes they bunch themselves up in a corner for several days and get parched and injure themselves. And as they don't know the terrain, they can't find the watering holes, the windmills, the dams or puddles, they have to be rounded up and led to the water. In that same way, for many days they get bunched up together in scattered parts of the horse pasture, so we have to spread them apart to let them find their own pasturage, and we have to inspect the pasture and watering holes as the time goes by to see how many steers are coming down to the water, whether for this windmill or the other, or at this dam here and another further ahead, to be sure that the steers are learning the terrain, that they don't suffer.

While making one of those rounds of the cow pastures, I arrived one afternoon at the dam of La Carabina so my horse and I could drink water, and then I went on foot all along the bank of the dam reservoir, looking to determine

novillos estavan más o menos caiendo al agua allí y fuí
notando que ha aquí bajavan dos o tres y más delante un
atajito chico y más álla desparramadose seña que ya se
estavan haciendo al potrero, y andavan desparramadose.

De repente noté una fuea desconocida entre las de un
atajito de novillas entre el soquete suave en la orilla del agua
y me pareció que era de potrillo mediano. Pero me pareció raro porque en ese potrero no había caballos ni lleguas
menos y tenía que venir de afuera. Bueno, me fuí siguiendo
el potrero y ya cuando pisó tierra humeda, más macizo,
entonces ya la fuea parecía más de burro que de potrillo,
pero ya pa' cuando pisó tierra maciza ya la fuea no era de
burros . . . era de mula.

Y yo estaba seguro de que en ese potrero no había mula
. . . tenía que benir de afuera. La seguí por un pedaso pero
ya cuando so revolvió con ha fuea de los novillos, la perdí y
ya pa' entonces estava oscureciendo, muy poca luz no había
chansa de seguirla.

Me fuí pa'l Rancho quo estava lejos de allí.

CUANDO LLEGUÉ, desensillé y le eché pastura a mi caballo
pa' que cenara, y luego me fuí pa'l campo. Allí estava el
caporal y mientras que servía una taza de cafe, le dijé, "Te
traigo unas nuevas. Anda una mula en el potrero. "

"Onde las viste?" me pregunto medio enojado.

"Pues, no la ví, pero le ví la fuea en la presa de La
Carabina."

"Pues, búscala," me dijó, "y echala fuera porque el patrón
es muy delicado. No quiere mulas en el potrero porque corren
los novillos porque se va a enojar mucho si la llega a ver él.
Mañana nomás amanece, vas y la buscas, y ha hechas fuera."

Pero, no le pusé mucho cuidado porque era más la
soflama de él y del patrón que ni daño que iva a ser la mula.
Si es cierto que ay mulas que les da por retosar y jugando,
agarran de correr al ganado. Pero, eso lo hacen las mulas
gordas, descarigadas, que no tienen más que hacer coma
algunos que les sobra su diriero y les gusta pasar un buen
tiempo y no les importa con quien, crellendo que el din-
ero paga todo. Y, esta pobre mula yo figuraba que había de

how many steers, more or less, were coming down to the
water. And I began to notice that two or three more were
already coming down ahead of a small herd and spreading
apart farther away, perhaps because they were settling into
the pasture, they were already scattered. Suddenly I noticed
one odd set of tracks among those of a small herd of steers
in the soft mud at the edge of the water. And it looked to me
like that it was of a small colt. But that seemed odd because
in that pasture, there were no horses much less foaled
mares. And it had to be from outside of the pasture. So I
went along the pasture, and when the tracks were in damp
dirt, more solid, then at that point they looked more like
tracks of a donkey than of a colt. But by the time it stepped
in solid dirt, now the tracks were not of a donkey . . . they
were mule tracks.

And I was sure that there was no mule in that pasture . . .
it had to have come from outside. I followed the trail for a
piece, but when it mixed in with the tracks of the steers, I
lost it. And by that time, it was getting dark. Very little light
gave me no chance of following it.

I went to the camp, which was far from there.

WHEN I ARRIVED, I unsaddled my horse and fed him hay
to eat, and then I went over to the camp. The foreman was
there, and as I served myself a cup of coffee, I told him that I
had some news. "There's a mule in the pasture."

Where did you see it?" he asked me half angrily.

"Well, I didn't see it, but I did see its tracks at the dam of
La Carabina."

"Well, look for it," he told me, "And get her out because
the boss is very sensitive, and he doesn't want mules in the
pasture because they chase the steers. And because he'll get
very angry if he comes and sees her. Tomorrow, at first light,
go and find her, and put her out."

But I didn't pay much attention because it was more a
squabble between him and the boss, and because the mule
wasn't going to do any harm. It is true that there are mules
that are given to romping about, and playing along, they
take to chasing the cattle. But that's done by fat mules that

estar en muy malas condiciones pa' que so huviera arro-
jado a meterse, brincando sercas esponiendose a cortarse o
matarse. Muy seguro que las necesidad las exigía el ambre
y quién save de donde vendría. Yo por lo pronto figuré que
era de algun ranchero chiquito que pegava con nosotros y
los más eran gentes que yo conocía bien. Pensé, un día que
los mire, les pregunto que si no se les ha perdido una mula.
Mientras que se vaya responiendo la pobre que bastante ha
de haber sufrido.

Otro día como era de costumbre en esos tiempos del
vaquero en esos tiempos, salí muy de mañana rumbo a la
presa de La Carabina a que me fuera manecer alla con la
idea do aver si podía ver la mula. En esos alrededores no con
idea de echarla fuera si nomás pa conocerla y ver que marca
tenía y podía ser que yo conociera el fierro y saver de quien
era su dueño.

Cuando llegué alla, al lugar donde se me había perdido
la fuea de la mula, ya estava claro, y traté de pescarle otra
vez pero no pude porque con su trío de novillos que habían
andado allí esa noche, no se mirava ninguna seña pa' que
rumbo había ganado pa' más o menos saver su comedero.

Entonces, me puse a buscarla a los alrededores. Podía ser
que la devisara de lejos porque era terreno de loma y planes.
O de malas, verle la fuea otra vez. Pero después de un buen
rato de andar campeando, no conseguí nada. Yo no me inte-
resaba mucho hallar la porbre mula, necesitaba la alluda,
había que darle chansa a que viviera.

 razón porque anduviera allí y
el sacate que se comiera en un potrero tan grande y tan
empastado no era daño, era como quien le quita un pelo a
un buey.

En la tarde que volví al campo, luego me preguntó el
caporal, "Eschaste fuera la mula, Juan?"

"No," le dije, "no la hallé, no la fuea siguiera."

"Siguela buscando."

"Ta bueno," le dije. Y se volvío canción de preguntar
por la mula todas las tardes que volvía del potrero. Y era la
misma respuesta hasta que se cansó.

are rested and have nothing more to do, like some people who
have extra money and want to have a good time with little
regard to who with, thinking that money can buy anything.
And this poor mule, I figured that she must be in bad condition to have gahered in here, jumping over fences, exposing
herself to cutting herself or killing herself. Most surely, her
need was driven by hunger, and who knows where she came
from. And so I simply figured that she was from some small
ranch next to ours, and probably they were people that I
knew well. I thought, some day when I see them, I'll ask them
if they had not lost a mule. Meanwhile as the poor thing goes
along, that she must have suffered quite a bit.

On another day as it was custom in those days of the
vaquero of those times, I set out very early in the morning
toward the dam of La Carabina so as to arrive there by dawn
in the thought that I might see the mule in that surrounding
area, not with the idea of putting her out, but rather to look
her over and see which brand she had, thinking that I might
know the brand and know who she belonged to and who was
her owner.

When I arrived there, the place where I had lost the trail
of the mule, it was already daylight and I tried to get the
trail again but I could not because of a trio of steers that had
walked there that night, there was no sign in sight of which
direction she had taken to indicate more or less her grazing
ground.

Then, I set myself to find her in the surrounding area; perhaps I would be able to identify her from a distance because it
was a terrain of hills and plains, or at least see her trail again.
But after a good period of riding, I did not find anything. I
was not much interested in finding her. The poor mule, she
needed help, she needed a chance to live.

 any other reason to be there, and any grass
that she would eat in such a large pasture that was so grassy
that she couldn't do any harm. It'd be like cutting one hair on
an ox's tail.

That afternoon that I returned to the camp, right away the
caporal asked me, "Did you put out that mule, Juan?"

A los cuantos días y varias veces después, le volví a ver la
fuea en la presa donde bajaba al aqua y en el potrero pero
no muy frescas. Y noté que siempre andaba con un atajito
de novillos de compañeros. Parece algo estraño la compañía,
pero no es la soledad. Estar solo sin compañía ninguna es
una cosa terrible. No los animales la aguantan. Se secan los
huesos en vida. La pobre mula tenía que tener compañía,
desde luego que no había allí nadie de su qente allí, ni
caballos ni mulas. Yo he savido de diferentes casos de com-
pañías raras entre animales como un caballo solo tener de
compañeras a las vacas, un toro ladino y una mula ladina ser
intimos compañeros cuando nomás mas ellos dos era todo lo
quedaba en ese potrero, y lo más raro, una potranca ladina
y una vaca ladina de compañeros y cuando le echaban los
perros a la vaca pa' pescarla, la potranca peleaba con ellos a
rnanotadas pa' que la vaca se escapara.

Así es que con la curíosida de conocer la mula, siempre andaba con la precaución observando a ver si be podía
pescar la fuea fresca y a los novillos para seguirla por la fuea
tenía que dar con ella y sus compañeros porque era buen
atajito facil de hallarlos.

Pués un día de tantos, lliendo y viniendo, dí con las fueas
pocos frescas y me fuí siguiendolos. Los novillos y la mula
iban segun las fueas despacio, comiendo y caminando. Y mi
caballo al trote, no gaste mucho en alcansarlos en una abra
pero nomás los novillos . . . la mula no estaba allí.

Pués, me quedé parado viendolos y sorprendido porque la
fuea de la mula iba junto con ellos y era muy raso alrededor.
No me quedava ninguna duda que no estaba allí la mula y
como todo el interés mio era nomás conocerla, no la busque
más por ese día porque tenía que atender algo más delante.
Me conformé con decirme, "En otra ocasión la miro—la
conozco." Y seguí caminando pa' mi destino, sin la menor
sospecha o mala de en la mula, sino con mi corazón muy
tranquilo que estaba haciendo una obra de caridad con ella,
muy contento me fuí entonando una canción con la confianza
y satisfación que si la mula pudiese darse cuenta que yo le
estaba haciendo un favor . . . me lo agrecedería mucho.

Pasó tiempo, y no volvía ver la fuea de la mula y por lo
pronto pensé quo se había ido del potrero. Ya fuera al lugar

"No," I told him, "I didn't find her, at least not her trail."
"Keep searching her."

"Okay," I told him. And so it turned into a sing-song with
him asking me for the mule every afternoon that I returned
from the pasture, and with the same answer until he got
tired of asking.

After so many days and so many times afterward, I saw
again her tracks leading down to the water at the dam as
well as in the pasture, but not very fresh tracks. And I noted
that she was always with a small herd of steers as companions. It seemed a strange thing this company, but it's not the
solitude. Being alone with no companion at all is a terrible
thing. No animals can stand it. One's bones go dry in life.
The poor mule had to have company since there were none
of her kind around, neither horses nor mules. I have known
of different cases of companionships among animals like a
sole horse having companions among cows, a wild bull and a
wild mule being intimate companions when only those two
were all that was left in a pasture. And the most rare, a wild
colt mare and a wild cow as companions. And when I would
sic the dogs at the cow to catch her, the mare colt would
fight them off with her hoof so the cow could escape.

And so it was that with the curíosity of learning the
mule. I was always cautiously looking to see if I could catch
her fresh trail and that of the steers by her trail. Her trail
had to lead to her and her comanions because it was an easy
little herd to find.

Well, one of those days, after coming and going, I came
upon their fresh tracks, and I went following them. The
steers and the mule were going slowly judging by their
tracks, eating and walking. And my horse at a trot, it didn't
take me long to catch them in a clearing. But only the steers
. . . the mule was not there.

Well, I stood watching them puzzled because the trail of
the mule went along together with them, and the ground
was clean everywhere else. I had no doubt that the mule was
not there, and since my only purpose was to find her, I did
not search her further that day because I had other matters
to tend to ahead. I resigned myself by saying, "on another
occasion I see her—I find her." So, I went on my way to my

donde era su casa o que fuera de pasada al rumbo que llevaba porque lo que me ise creer eso fué que su atajito de novillos con quien andava, todavía andavan en los mismos comederos.

UN DÍA CAMPEANDO LLEGUÉ a un papalote y se había tirado mucha aqua de la canoga. Ya se había echo un charco grande y me puse a arreglar el tambor que no tapava bién. Luego, volví a montar a caballo. Me fuí al paso fijandome alrededor a ver cuantos novillos estavan bajando a ese aguage.

Y cuando menos esperaba, vi la fuea de la mula entre la de los novillos. Y me sorprendió que andaba hay— ciendo aca tan lejos de donde la había visto primero con otro atajo. Traté de seguirla pero no estava muy fresca. No había chansa de alcansarlo. Como quiera, la seguí un pedaso. Nomás pa' saber, más o menos, que rumbo llevaba, para otro día que tuviera más tiempo buscarlo por esos alrededores.

Como no había de ser paso, a los cuantos días que pasé por allí, me toco la buena suerte de pescar ha fuea de a mula poco fresca con el atajito de novillos. Y me puse a seguirla ahora si había chansa de conocerla, por el tiempo que ensaeñaba ha fuea de aver pasado.

No podía estar muy lejos porque por lo regular, cuando salen del agua no se van muy recio y se van entreteniendo. Ya sea que ban comiendo aquí y alla, o que so entretangan con otros novillos amigos que ban rumba al aqua, como quien se para a platicar con su amigo que se encuentra en su mismo camino.

Y así fué como yo pensaba. Al no mucho rato, los alcancé a devisar de lejos de una loma a otra. Pero no se alcansaban a devisar todos muy bién porque había poco monte. Pero, ya cuando crucé redamadero que había de por medio, entonces si los pesque de serca y los pude ver bién.

Pero la mula no estaba con ellos. Y anduve buascando rededor y fijandome bién porque era fácil de ver porque el monte estaba raso. Y al mismo tiempo, fuí cortando fuea a ver pa' que rumbo había ganado, pero no pude ver nada. Pero no podía estar muy seguro de que no le huviera errado porque en terreno de loma y en tiempo seco, no es muy fácil destinguir la fuea de una animal solo.

destination, with no bad suspicion or evil intent toward the mule, but rather with my heart very satisfied that I was doing a work of charity by her, very contently humming a tune as I went along with the confidence and satisfaction that if the mule were able to recognize that I was doing her a favor . . . she'd be very grateful to me.

Time passed, and I did not again see the trail of the mule. With that, I thought that she must have left this pasture, whether to her home pasture or that she had just gone along on her way. What made me think that was that the little herd of steers that she had traveled with, were still grazing in these same pastures.

RIDING ALONG ONE DAY, I arrived at a windmill. A lot of water had spilled from the watering trough and made a large puddle, so I set about to repair the float that wasn't closing well. Then, I went back and mounted my horse and went at a good pace, looking around to see how many steers were going down to that drinking spot.

And when I least expected it, I saw the tracks of the mule in among those of the steers. It puzzled me that she would be doing over here so far from where I had first seen her with another herd of cattle. I tried to follow the trail, but it wasn't fresh. There was no chance of following it. Anyway, I followed it a piece. Just to know, more or less, which direction it was going so some other day when I had more time, I could search it around that area.

Since there was no other way around, a few days later that I passed through there. I had the good fortune of spotting the mule's trail, quite fresh, and going along with the small herd of steers. So, I set about to follow it to see if now I could find her, the trail showing the time that has passed. It couldn't be too far because normally when they leave the water, they don't go very fast, and they go along entertaining themselves. Either because they are grazing here and there, or that they entertain themselves with other steer friends who are coming this way to the water, like someone who stops to chat with their friend that they meet coming along the same road.

Cualesquiera se equivoca, y como para mi no era de mucho importancía hallarla, no persistí mucho en buscar la mula. Desde luego que mi intención no era echarla fuera. Me fuí pa'l rumbo que llebava campeando, revisando otros novillos, aguages, y cercas.

Era muy grande el terreno . . . Muy larga la vuelta varia potreros.

And it was as I had thought, as I had figured, that not too much later, I caught a glimpse of them at a distance from one hill to the other, although not all of them were very well visible because there was some brush. But after I crossed a thicket along the way, then I did come close upon them where I could see them well, but the mule was not with them.

So, I went looking around, noting carefully because it was easy to see since the brush was sparse. At the same time, cross-tracking the trail to determine the direction it went, but I couldn't see anything.

But I couldn't be too sure that I hadn't missed her because, alone in hilly terrain when it's dry, it's not very easy to distinguish one animal's tracks, and anybody can get confused. And since for me, it was of no great importance to find her, I did not persist very much in finding that mule. Anyway, it was not my intention to put her out. I went riding along on my way, checking other steers, watering holes, and fences.

It was very vast, the open range . . . and so very winding, the trail through horse pastures.

Requiem

As a flare in the ashes,
A smile among the dead,
That is the great spirit of the brave,
Of the free,

High above the darkest hour,
He rises,
Blazing the trail of the most sacred heritage.
It is a symbol in sunlight,
As a crowning of the blesssed,

It is an emblem of the heavenly light
In gold leaf,
The God given gift to stand free
And not to crumble in despair.

Ricardo Moreno Beasley
Wednesday, August 14, 1974
Alice, Texas

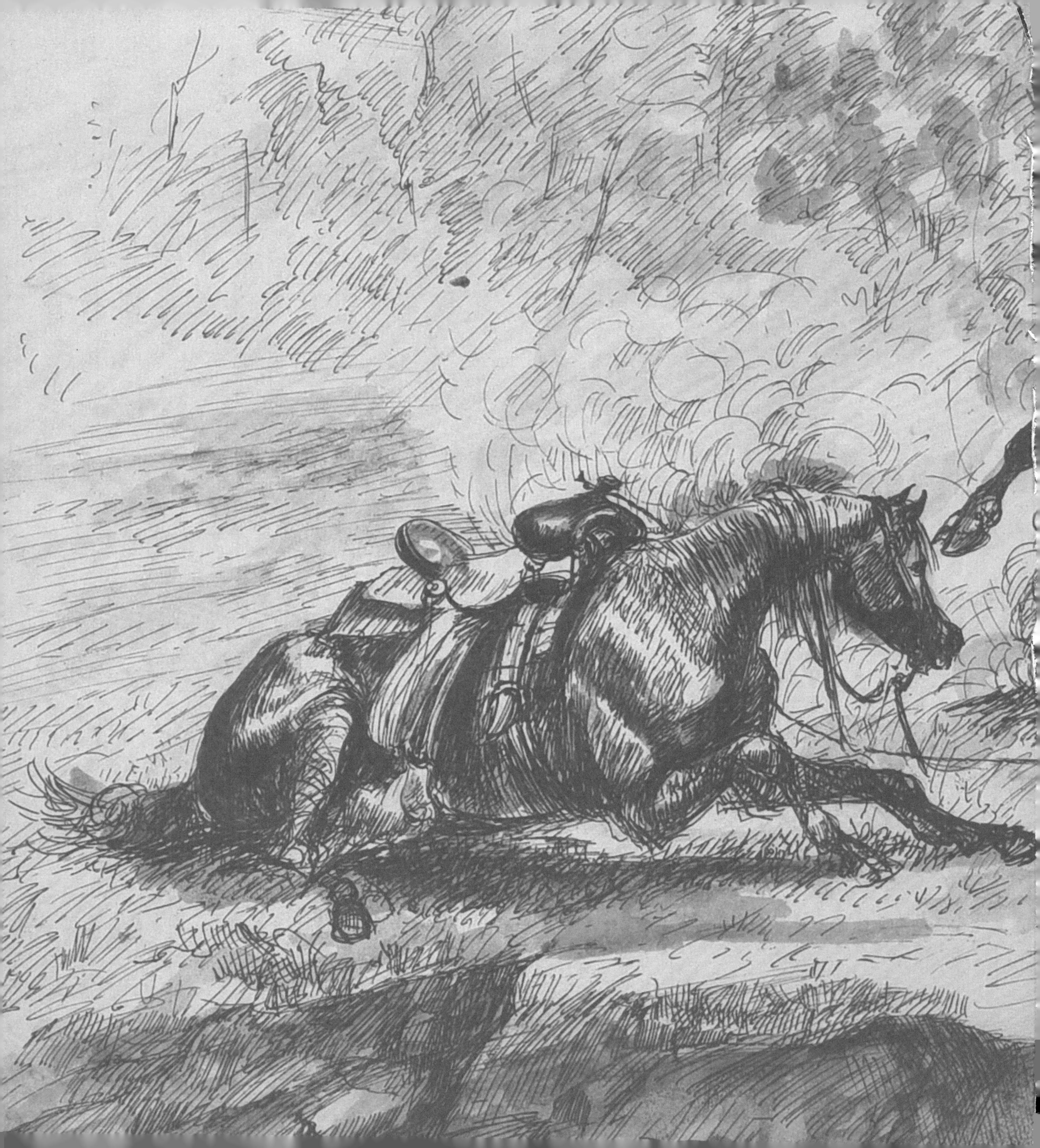